Equilibrium Models and Variational Inequalities

This is volume 210 in
MATHEMATICS IN SCIENCE AND ENGINEERING
Edited by C.K. Chui, *Stanford University*

A list of recent titles in this series appears at the end of this volume.

Equilibrium Models and Variational Inequalities

I.V. Konnov
DEPARTMENT OF APPLIED MATHEMATICS
KAZAN STATE UNIVERSITY
KAZAN
RUSSIA

ELSEVIER
Amsterdam – Boston – Heidelberg – London – New York – Oxford
Paris – San Diego – San Francisco – Singapore – Sydney – Tokyo

Elsevier
Radarweg 29, PO Box 211, 1000 AE Amsterdam, The Netherlands
The Boulevard, Langford Lane, Kidlington, Oxford OX5 1GB, UK

First edition 2007

Library of Congress Cataloging-in-Publication Data
A catalog record for this book is available from the Library of Congress

British Library Cataloguing in Publication Data
A catalogue record for this book is available from the British Library

ISBN-13: 978-0-444-53030-1
ISBN-10: 0-444-53030-4
ISSN: 0076-5392

Transferred to Digital Printing 2007

Preface

The concept of equilibrium plays a central role in various applied sciences, such as physics (especially, mechanics), chemistry, biology. In terms of the classical mechanics, for example, the equilibrium state for a system means that

(a) *the impact of all the forces on this system equals zero*;

and

(b) *this state can be maintained for an indefinitely long period.*

Thus, one can formulate the equilibrium problem mathematically, i.e. in the form of a mathematical model, and the solutions of the corresponding problem can be used for forecasting the future behavior of the system and, also, for correcting the deviation between the current state of the system and the equilibrium state. The standard mathematical model for the equilibrium state is a system of equations, which often admits finding a solution in an explicit (closed) form. Observe that solution sets of most of these problems possess certain vector space properties or represent its regular transformation, such as a manifold, and their analysis relies upon these properties. Even the classical equilibrium concepts enable us to solve many difficult problems in natural applied sciences. However, the necessity of investigation of more complicated problems, which are essentially nonlinear ones, requires generalizations of these concepts. Of course, the equilibrium approach has certain restrictions, i.e. it is not applicable to arbitrary systems. For example, systems with rapidly changing states should be investigated with the help of some other tools. However, in each field we can observe many systems which exist without essential changes for a long time or admit some stable modifications in their state. It is very essential to note that the generalized concept of equilibrium is not restricted by the static problems only; hence, one can consider the equilibrium trajectories which correspond to dynamic equilibrium. Such systems can be investigated via just the generalized equilibrium approach, but now one can not expect the derivation of explicit solutions or manifold type properties, so that this will also require new mathematical models and methods.

We spoke about the situation in natural sciences, but in the socio-economical sciences the equilibrium approach may be even more powerful and fruitful. The point is that these sciences, in contrast to the natural ones, do not admit in fact other kinds of modeling with the exception of the mathematical one. Hence, suitable formulations of equilibrium models enable us to make non-trivial conclusions on the behavior of very complicated systems which are considered in socio-economical sciences. Thus, the *first key problem is to find suitable formulations of equilibrium models in these fields.* It should be noted that equilibrium models were very developed traditionally in economics. Many Nobel Prize winners, such as *K.J. Arrow, G. Debreu, L.V. Kantorovich, T. Koopmans, W. Leontief, H. Markovitz, J.F. Nash et al.*, were awarded just for their contributions in this field. Nevertheless, even in economics, there exist a number of various kinds of equilibrium models, even different concepts of equilibrium. These models are investigated and applied separately and for this reason they require different tools both for the theory and for the construction of solution methods. This course can be regarded as an attempt to present a unifying look on equilibrium concepts in economics. Such an approach will require certain extensions of the usual concept of equilibrium and a presentation of unifying tools for investigating and solving equilibrium models which may have in principle very different initial sources. Moreover, it forces us to give simplified versions of many results and methods and to drop some details in particular models which however may be very interesting and deep within these models. At the same time, we include several models from related sciences which demonstrate very broad fields of possible applications of the equilibrium approach.

Of course, together with formulations of equilibrium problems, there are many very essential issues to be investigated and resolved, such as the *existence and uniqueness of solutions, stability of equilibria, existence of suitable and transparent mechanisms for attaining equilibrium states,* and *creating effective algorithms for finding equilibrium solutions.* We also discuss them in this book, but also in connection with the unified framework of equilibrium models. For this reason, we describe mostly methods which rely upon the generic properties revealed in the models. Observe that the above questions are usually investigated and answered separately for each particular model and field of applications. Therefore, existence of such a unified system of equilibrium models yields a great support for creating successful and effective tools for resolving all the basic questions.

This book is based upon the lecture courses presented by the author at Kazan University, Bergamo University, and Oulu University. It involves some results obtained within the scientific work at Kazan University and Informatics Problems Institute, AS RT.

I would like to express my sincere gratitude to all the people who helped me in preparing this book and sent me their comments and opinions. I am especially grateful to E. Allevi, M. Bertocchi, and E. Laitinen for the opportunity to present lectures at Bergamo University and Oulu University. I am also thankful to the Elsevier staff for their friendly co-operation.

Igor Konnov *Kazan*
May 2, 2006

Contents

List of Figures

Chapter 1

Introduction

We intend to consider equilibrium concepts and their applications in economics and related fields describing very complicated systems. Nevertheless, even very simple models can give a non-trivial information about a system under consideration if they take into account its essential features. In order to illustrate our approach to the development of equilibrium concepts, we start from the simplest one-dimensional models.

Let us consider a market where consumers and producers buy and sell, respectively, a homogeneous commodity, their reaction depending on the current commodity price, i.e. it is a completely aggregated model. More precisely, given a price p, the consumers determine their total demand $D(p)$ and the producers determine their total supply $S(p)$, so that the excess demand of the market is then the following

$$E(p) = D(p) - S(p).$$

Clearly, if we consider certain amount of transactions between consumers and producers then there exists the equality between the partial supply and demand at each price level, but the problem is to find the price which implies the equality between the total supply and demand, i.e. when

$$E(p^*) = 0. \tag{1.1}$$

It is called an *equilibrium price* and corresponds to the classical static equilibrium concept, where the impact of all the forces equals zero, i.e. it is the same as in mechanics. Moreover, this price implies constant clearing of the market and may be maintained for an indefinitely long period. This means that we are able to describe the behavior of the market and its economic agents. The first problem is to find conditions ensuring the existence of equilibrium prices, i.e. solutions of equation (1.1). However, such a classical formulation of the problem is not complete since it does not take into account the possible restrictions on prices. First of all, formulation (1.1) assumes tacitly the absolute flexibility of price, but it must be non-negative, and we obtain the non-classical problem

$$p^* \geq 0, E(p^*) = 0, \tag{1.2}$$

instead of (1.1). Next, many real markets may include two-side price rigidities, and we should then replace (1.1) by the following problem

$$p' \leq p^* \leq p'', E(p^*) = 0. \tag{1.3}$$

Of course, we can begin our considerations from the simplest case (1.1), where the desired conditions may be derived from the well-known *Cauchy* theorem. That is, if there exists a segment $[a, b]$ such that

$$E(a)E(b) \leq 0$$

and E is continuous, then there exists a solution of equation (1.1), or equivalently, the mapping $p \mapsto T(p)$ with $T(p) = p + \lambda E(p), \lambda > 0$, has a fixed point on the segment $[a, b]$, i.e.

$$p^* = T(p^*). \tag{1.4}$$

By imposing certain assumptions on the behavior of consumers and producers, we can ensure the existence of such a segment. Usually, these conditions are deduced from monotonicity properties of the functions (mappings) D and S on their domain or at least on some subsets containing the boundary of the feasible set. The classical conditions say that the demand tends to infinity and the supply tends to zero if the price drives to zero, conversely, the demand tends to zero and the supply tends to infinity if the price drives to infinity, i.e.

$$\begin{cases} D(p) \to +\infty & \text{and } S(p) \to 0 & \text{if } p \to 0, \\ D(p) \to 0 & \text{and } S(p) \to +\infty & \text{if } p \to +\infty. \end{cases} \tag{1.5}$$

Clearly, if we choose a close to zero and b large enough, (1.5) implies $E(a) > 0$ and $E(b) < 0$, i.e. both (1.1) and (1.2) have solutions, moreover, (1.5) shows that we can neglect the non-negativity constraint and simply consider only the classical equation (1.1); see Figure 1.1.

However, (1.5) may not hold for many real markets due to the boundedness of both supply and demand, moreover, they may be bounded away from zero as well, which does not ensure the existence of a solution, at least for problem (1.2); see Figure 1.2.

Of course, in order to establish the existence of solutions for problem (1.3), we have to overcome even more essential difficulties, since in general

Figure 1.1:

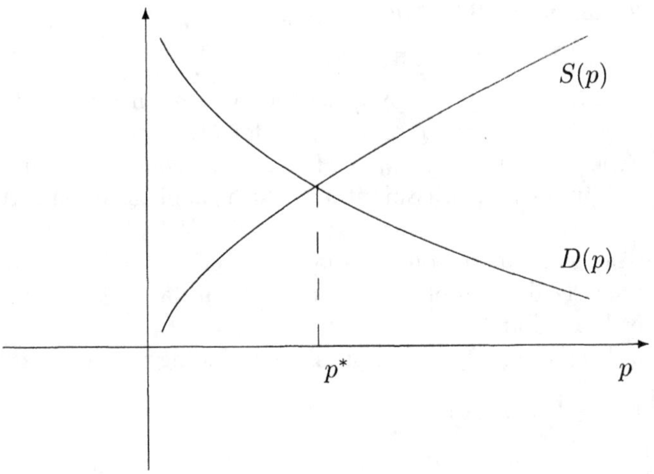

Figure 1.2:

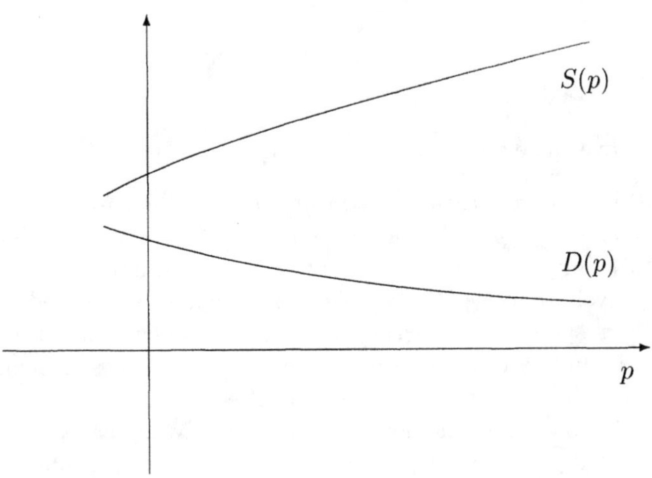

we have to find the corresponding segment $[a, b]$ inscribed in $[p', p'']$, or equivalently, to find $\lambda > 0$ and the segment $[a, b] \subset [p', p'']$ such that T maps $[a, b]$ onto itself for the case (1.4). Observe that $-D$ and S are supposed to usually possess global monotonicity properties, i.e.

$$D(p_1) \geq D(p_2) \text{ and } S(p_1) \leq S(p_2) \text{ if } p_1 \leq p_2,$$

but they do not ensure existence of solutions for (1.2) or (1.3) in general, as Figure 1.2 illustrates; although somewhat strengthened properties may in principle yield the existence of solutions for (1.2); see Figure 1.1. Clearly, the existence of equilibrium prices for market models with many commodities will require stronger assumptions which enable us to omit the restrictions and to consider only classical equations like (1.1), but they may appear non-realistic. For this reason, we should present extended equilibrium concepts, which lead to problems whose solvability may be obtained under comparatively mild and natural assumptions.

Namely, we will replace (1.2) by the problem of finding p^* such that

$$p^* \geq 0, E(p^*) \leq 0, p^* E(p^*) = 0; \tag{1.6}$$

which is called *the complementarity problem* with the cost mapping $-E$. Note that its solution satisfies (1.2) if $p^* > 0$, but $p^* = 0$ in (1.6) only implies the non-positivity of excess demand at 0. Similarly, (1.3) can be replaced by the following problem: Find $p^* \in [p', p'']$ such that

$$E(p^*) \begin{cases} \leq 0 & \text{if } p^* = p', \\ = 0 & \text{if } p^* \in (p', p''), \\ \geq 0 & \text{if } p^* = p''; \end{cases} \tag{1.7}$$

or equivalently,

$$E(p^*)(p^* - p) \geq 0 \quad \forall p \in [p', p'']. \tag{1.8}$$

Exercise 1.1. Prove the equivalence of (1.7) and (1.8).

Problem (1.8) is called *the variational inequality* with the cost mapping $-E$ and the feasible set $[p', p'']$. Again, any solution p^* of (1.7) may differ from that of (1.3) only if either $p^* = p'$ or $p^* = p''$, where only non-positivity (respectively, non-negativity) of excess demand is required. It means that the corresponding restrictions prevent for the current price to pass through the boundary of the feasible set although the excess demand is non-zero, and we obtain an extended concept of equilibrium. In fact, if $-D$ and S are monotone, then the decrease (respectively, increase) of the current price at p' (respectively, at p'') would drive the excess demand to zero. Observe also that (1.7) (or (1.8)) becomes equivalent to (1.6) if $p' = 0$

and $p'' = +\infty$. At the same time, (1.6) and (1.7) are solvable under much more general conditions than those in (1.2) and (1.3), respectively.

For instance, suppose that E is continuous. Then (1.6) is solvable if there exists a point $\tilde{p} \geq 0$ such that $E(\tilde{p}) \leq 0$. In fact, $E(0) \leq 0$ implies that 0 solves (1.6), otherwise we have $E(0) > 0$, i.e. there is a point $p^* \in [0, \tilde{p}]$ such that $E(p^*) = 0$. Similarly, we can derive simple solvability conditions for problem (1.7).

Exercise 1.2. Suppose that E is continuous and prove that (1.7) is solvable if $-\infty < p' \leq p'' < +\infty$. Find sufficient solvability conditions for (1.7) if either $p' = -\infty$ or $p'' = +\infty$.

This approach to formulating the static equilibrium admits further extensions for more complicated systems where their structure is taken into account explicitly or implicitly and may yield additional relations in the model. For instance, if we consider a set of regional markets joined by a transportation network, capacities of nodes and roads and shipment costs are taken into account and lead to some other equilibrium conditions, which do not ensure the pure price equilibrium for each separate market.

Let us now turn to dynamical processes corresponding to the static equilibrium of forms (1.1)–(1.3) or to their extensions given in (1.6)–(1.7). The role of dynamical processes is two-fold. On the one hand, they describe real changes in the system and we can investigate stability issues of these processes reflecting the same properties of the initial system. On the other hand, stable and rapid processes may serve as bases for computation of equilibrium points. Also, we may utilize both continuous and discrete time models for these purposes.

For simplicity, we consider the model with affine functions of demand and supply, i.e. let

$$D(p) = \alpha - \beta p, S(p) = \gamma + \delta p \text{ where } \alpha, \beta, \gamma, \delta > 0. \tag{1.9}$$

Then $-D$ and S are clearly (strictly) monotone and (1.1) yields the explicit formula for the equilibrium price

$$p^* = \frac{\alpha - \gamma}{\beta + \delta}, \tag{1.10}$$

i.e. p^* is defined uniquely and positive if $\alpha > \gamma$. It means that the initial value of demand at zero has to be greater than the same value of supply. Then we can consider the well-known "cobweb" process

$$D(p_k) = S(p_{k-1}), \; k = 0, 1, \ldots, \tag{1.11}$$

Figure 1.3:

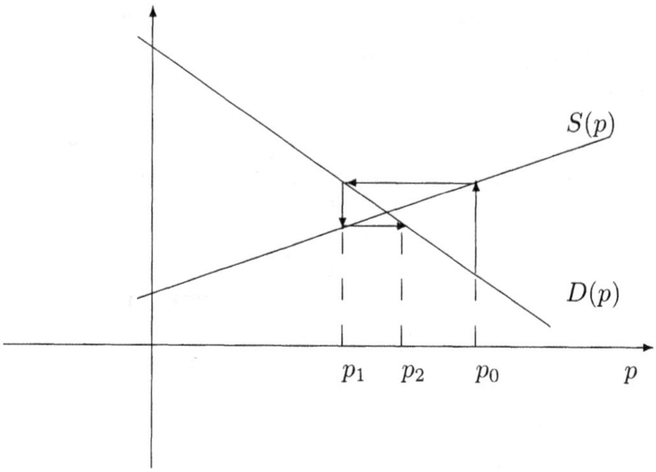

for modeling behavior of prices or the negotiation procedure, where p_k denotes the price at the k-th time segment. On account of (1.9), we have

$$p_k = \frac{\alpha - \gamma}{\beta} - \frac{\delta}{\beta} p_{k-1}, \; k = 0, 1, \dots, \tag{1.12}$$

hence

$$p_k - p_{k-1} = \frac{\delta}{\beta}(p_{k-2} - p_{k-1})$$

and we obtain the clear global criterion of convergence:

$$\delta/\beta < 1, \tag{1.13}$$

i.e. the growth ratio of supply has to be less than that of negative demand. In fact, (1.13) implies $|p_k - p_{k-1}| \to 0$ as $k \to \infty$ and (1.12) now yields $p' = p^*$ for the limit point p' of the sequence $\{p_k\}$. Thus, condition (1.13) ensures stable and rather rapid convergence of process (1.12) to the equilibrium point regardless of the choice of the starting price p_0; see Figure 1.3. Observe that process (1.11) has rather clear interpretation and means that the current prices result from the equality of demand and supply, but in different time periods. Namely, the price adjustment is immediate for demand, but has one period delay for supply. We should draw our attention to the fact that neither existence of the unique equilibrium point nor

Figure 1.4:

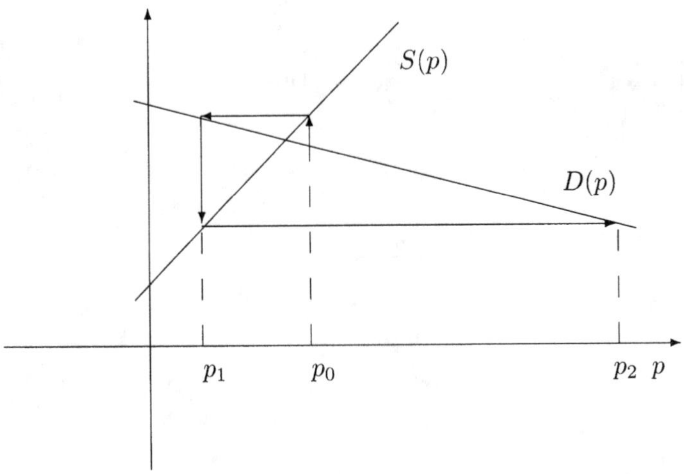

strengthened monotonicity properties of $-D$ and S ensure the convergence of the dynamical process (1.11) if condition (1.13) is not satisfied. In fact, the reverse condition

$$\delta/\beta > 1 \tag{1.14}$$

leads to the divergent sequence $\{p_k\}$ for an arbitrary starting point; see Figure 1.4. Also, the equality $\delta = \beta$ gives the cyclic procedure $p_{k+1} = p_{k-1}$ for $k = 1, 2, \ldots$, i.e. the behavior of the process is stable, but it is not convergent. Observe that the "reverse" process to (1.11) is written as follows

$$D(p_{k-1}) = S(p_k), \ k = 0, 1, \ldots,$$

and, due to (1.9), has the reverse convergence condition (1.14), whereas (1.13) yields its divergence and the equality $\delta = \beta$ yields the same cyclic property.

From this simple analysis we can conclude that stable convergence to an equilibrium state requires additional conditions to those ensuring the existence of such equilibrium states, i.e. the existence and even uniqueness of equilibrium do not ensure convergence. At the same time, we can even interpret any dynamic process with continuous functions as stable in the sense that there exists a fixed formula joining system states in different time periods. For instance, (1.12) describes the stable relation between p_k

Figure 1.5:

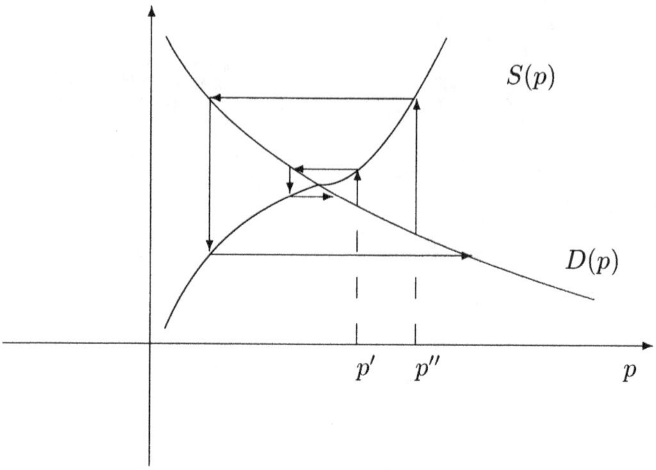

and p_{k-1} although their values may vary and their behavior may change for different relations between δ and β, i.e. (1.12) is a stable dynamic process in this extended sense.

If we consider nonlinear functions of demand and supply as in Figure 1.1, we obtain the same conclusions on the behavior of the cobweb process (1.11) in general, it may be additionally dependent of the choice of the starting price p_0, as Figure 1.5 illustrates. In fact, the process with starting point p' converges, but this is not the case for the starting point p''.

This is not so easy to give a suitable continuous formulation for process (1.11). Let us consider some other dynamic process which admits both formulations. Its continuous form is the following:

$$\frac{dp(t)}{dt} = E(p(t)), \ p(0) = p_0,$$

and we obtain the discrete form similarly

$$p_{k+1} = p_k + \lambda_k E(p_k), \lambda_k > 0, k = 0, 1, \ldots$$

That is the price increases (respectively, decreases) if the current value of excess demand is positive (respectively, negative). This process also seems very natural.

In the affine case, we have

$$\frac{dp(t)}{dt} = (\alpha - \gamma) - (\beta + \delta)p(t), p(0) = p_0,$$

and

$$p_{k+1} = p_k + \lambda_k \left[(\alpha - \gamma) - (\beta + \delta)p_k\right], \lambda_k > 0, k = 0, 1, \dots,$$

respectively. The continuous trajectory may be determined explicitly:

$$p(t) = (p_0 - p^*)e^{-(\beta+\delta)t} + p^*$$

and has the unique stable stationary point p^* from (1.10). However, the discrete process converges to the solution p^* if the stepsize λ_k is small enough.

Of course, the analysis in the case of many commodities and in the presence of additional restrictions is more complicated but it in general leads to similar conclusions, where monotonicity type properties play the crucial role. The task is to formulate a suitable general equilibrium problem and to reveal properties in particular models, which are useful for analysis of this problem.

One of the most suitable and general formats for investigating and solving various equilibrium models is known to be the variational inequality problem. The single-valued *variational inequality problem* (VI) is the problem of finding a point $x^* \in X$ such that

$$(x - x^*)^T G(x^*) \geq 0 \quad \forall x \in X, \tag{1.15}$$

where X is a nonempty convex set of a Euclidean space E, $G : X \to E$ is a given mapping; cf. (1.8). Here and below, all the vectors are column ones, the superscript ^{T} denotes transpose, i.e. a^T denotes a row vector. It means that $a^T b = \langle a, b \rangle$ denotes the inner (or scalar) product of vectors. If we take E to be the n-dimensional Euclidean space $\mathbb{R}^n$, then

$$a^T b = \sum_{i=1}^{n} a_i b_i.$$

We recall that the set X is said to be *convex* if for each pair of points $x, y \in X$, the segment $[x, y]$ is contained in X, i.e. $\alpha x + (1 - \alpha)y \in X$ for all $\alpha \in [0, 1]$. Also, the set X is said to be a *cone* if $x \in X$ implies $\alpha x \in X$ for all $\alpha > 0$.

VI's are closely related with many general problems of Nonlinear Analysis, such as complementarity, fixed point and optimization problems. The simplest example of VI is the problem of solving a *system of equations.* It is easy to see that if $X = E$ in (1.15), then VI (1.15) is equivalent to the problem of finding a point $x^* \in E$ such that

$$G(x^*) = 0.$$

If the mapping G is affine, i.e., $G(x) = Ax - b$, then the above problem is equivalent to the classical *system of linear equations*

$$Ax^* = b.$$

Let us consider the case when X is a convex cone in E. Then VI (1.15) is equivalent to the *complementarity problem* (CP for short):

$$x^* \in X, G(x^*) \in X', (x^*)^T G(x^*) = 0, \tag{1.16}$$

where $X' = \{z \in E \mid x^T z \geq 0 \quad \forall x \in X\}$ is the dual cone to X. For instance, if we set E to be $\mathbb{R}^n$ and X to be the non-negative orthant $\mathbb{R}^n_+ = \{x \in \mathbb{R}^n \mid x_i \geq 0 \quad \forall i = 1, \ldots, n\}$ in $\mathbb{R}^n$, CP(1.16) can be rewritten in the standard form:

$$x_i^* \geq 0, \; G_i(x^*) \geq 0, \; x_i^* G_i(x^*) = 0 \quad \forall i = 1, \ldots, n;$$

cf. (1.6). The *linear complementarity problem* (LCP for short) corresponds to the case when G is affine, i.e., $G(x) = Ax - b$.

Next, if the mapping G is defined by $G(x) = x - T(x)$, where T maps X into itself, then problem (1.15) coincides with the *fixed point problem*: Find a point $x^* \in X$ such that

$$x^* = T(x^*).$$

Moreover, if the mapping G is the gradientmap of a real-valued function $f : X \to \mathbb{R}$, then problem (1.15) represents a necessary condition of optimality for the following *optimization problem* of finding a point $x^* \in X$ such that

$$f(x^*) \leq f(x) \quad \forall x \in X,$$

or briefly,

$$\min \to \{f(x) \mid x \in X\}.$$

Also, if the function f is convex, then the reverse assertion is true. Thus, all these problems can be viewed as particular cases of VI. Similarly, we can define VI with multi-valued cost mapping G and its related problems.

The theory and solution methods for various kinds of VIs are developed rather well and allow one to choose a suitable way to investigate each particular problem under consideration. So, the problem is to show that many applied problems can be easily formulated as VIs or their subproblems.

Part I

MODELS

In this part, we describe several different classes of equilibrium models and their mathematical formulations as optimization, complementarity and variational inequality problems. We show that revealing essential features of the initial economic model and utilizing this additional information may give non-trivial and deep results in comparison with those in the general case. At the same time, the derivation of such results for most non-linear models requires the corresponding theoretical background and will be given in the next parts.

Chapter 2

Linear Models in Economics

We recall that input-output analysis is the study of qualitative relations between the output levels of the various sectors of an economy, which is used for national accounting and planning. *W. Leontief* presented input-output analysis in application to the United States economy (see *Leontief* (1966)).

Following this approach, the economy is divided into n production sectors, each of them produces a homogeneous commodity. For a fixed time period, if the i-th sector produces v_i units of the i-th commodity, then the j-th sector uses y_{ij} units for its own production and y_i units are used for final demand. Hence, we can present the simple balance table for the time period indicated for each commodity:

$$v_i = \sum_{j=1}^{n} y_{ij} + y_i \quad \text{for} \quad i = 1, \ldots, n.$$

Dividing the inputs by output, we obtain the so-called input-output coefficients

$$a_{ij} = \frac{y_{ij}}{v_j},$$

which indicate the amount of the i-th commodity for production of one unit of the j-th commodity. The key assertion of the input-output analysis is that *these coefficients are constant* that tacitly assumes the absence of technological revolutions in the economy. Then, these coefficients can be used for forecasting and planning in the next time period.

2.1 Open input-output model

First we consider the so-called *open input-output model* which considers the question of satisfying a given final demand in the next time period.

Suppose that the values of the final demand for each commodity are known and represented by the vector $y = (y_1, \ldots, y_n)^T$, and that the coefficients a_{ij} are constant. The problem is to find output values $x = (x_1, \ldots, x_n)^T$ which satisfy this final demand. In other words, the problem is to find $x \in \mathbb{R}^n$ such that

$$x_i - \sum_{j=1}^{n} a_{ij}x_j = y_i \quad \text{and} \quad x_i \geq 0 \quad \text{for} \quad i = 1, \ldots, n. \tag{2.1}$$

This problem can be rewritten equivalently as

$$(I - A)x = y, \quad x \geq 0, \tag{2.2}$$

where I is the $n \times n$ unit matrix, A is the $n \times n$ matrix with the entries a_{ij}. Thus, we have obtained a somewhat unusual mathematical problem, since it involves the classical system of linear equations and the nonnegativity constraints.

Observe that the linear equations in (2.1) can be regarded as equilibrium conditions between the supply x_i and the demand $\sum_{j=1}^{n} a_{ij}x_j + y_i$ (which involves the production and final demand) for each i-th commodity. In order to solve problem (2.1) we first consider this system of linear equations without nonnegativity constraints. The existence of a solution for the arbitrary final demand vector y is guaranteed if the matrix $(I - A)$ is invertible. Then

$$x = (I - A)^{-1}\, y. \tag{2.3}$$

Observe that

$$(I - A)\left(I + A + A^2 + \ldots + A^{k-1}\right) = I - A^k$$

for each k. If

$$\lim_{k\to\infty} A^k = \Theta, \tag{2.4}$$

where Θ denotes the $n \times n$ zero matrix, then

$$\lim_{k\to\infty} \left[(I - A)\sum_{i=0}^{k-1} A^i\right] = \lim_{k\to\infty} \left(I - A^k\right) = I,$$

i.e., by definition,

$$(I - A)^{-1} = \sum_{k=0}^{\infty} A^k. \tag{2.5}$$

Next, we recall that A is the matrix of input-output coefficients so that in the standard situation we can suppose that they are nonnegative. Hence, each its degree A^k is also nonnegative and so is $(I - A)^{-1}$ due to (2.5). Thus, if there exists a solution to the system

$$(I - A)x = y$$

for each nonnegative right-hand side vector y, it has to be nonnegative, i.e. it solves also the initial problem (2.2). Thus, using the special property of the input-output matrix A, we have shown that the conditions $x \geq 0$ are superfluous in (2.2).

The question whether condition (2.4) is satisfied or not can be answered by using the well-known *Perron-Frobenius* theory of nonnegative matrices.

Theorem 2.1. *(Perron-Frobenius; e.g. see Nikaido (1968), Chapter 2, Theorem 7.1). If A is an nonnegative $n \times n$ matrix, then*

(i) it has a nonnegative eigenvalue λ_A with a nonnegative eigenvector, such that $\lambda_A \geq |\mu|$ for any eigenvalue μ of A;

(ii) the matrix $(I - A)^{-1}$ exists and is nonnegative if and only if $\lambda_A < 1$.

Observe that (ii) follows from (i) and (2.4). This property can be replaced with a somewhat simpler condition.

Theorem 2.2. *(Hawkins-Simon; e.g. see Nikaido (1968), Chapter 2, Theorem 6.1). If A is an nonnegative $n \times n$ matrix, then the matrix $(I - A)^{-1}$ exists and is nonnegative if and only if the n sequential principal minors of $(I - A)$ are positive.*

Let us now consider simple examples of such matrices.

Example 2.1. Set $n = 2$, then both the conditions in Theorems 2.1 (ii) and 2.2 are not satisfied for the matrix

$$A = \begin{pmatrix} 1 & 1 \\ 1 & 1 \end{pmatrix},$$

but they are satisfied for the matrix

$$A = \begin{pmatrix} 1/2 & 1/2 \\ 1/3 & 1/3 \end{pmatrix}.$$

Exercise 2.1. Check the conditions of Theorems 2.1 and 2.2 for the matrix

$$A = \begin{pmatrix} 1/2 & 1/3 \\ 1/3 & 1/2 \end{pmatrix}.$$

However, the necessary and sufficient conditions of productivity given in Theorems 2.1 and 2.2 seem too complicated for their verification in economic applications, especially due to the usual high dimensionality of these problems. Therefore, we have to suggest some other conditions which are only sufficient ones, but more suitable for utilization. In fact, since a_{ij} are input-output coefficients, it is natural to suppose that $a_{ij} \leq 1$. But this condition is not sufficient as the following example illustrates.

Example 2.2. Set $n = 2$ and

$$A = \begin{pmatrix} 1/2 & 1/2 \\ 1/2 & 1/2 \end{pmatrix},$$

then any condition in Theorems 2.1 (ii) and 2.2 is not satisfied.

Let us consider somewhat strengthened conditions:

$$\sum_{j=1}^{n} a_{ij} < 1 \quad \text{for} \quad i = 1, \dots, n \tag{2.6}$$

and

$$\sum_{i=1}^{n} a_{ij} < 1 \quad \text{for} \quad j = 1, \dots, n. \tag{2.7}$$

Condition (2.6) means that the amount of each commodity which is used for production of one unit of all the commodities is less than one unit, whereas condition (2.7) means that the total amount of commodities which is used for production of one unit of each commodity is also less than one unit. These properties show that there are opportunities for creating inventories and satisfying the final demand.

Theorem 2.3. *Suppose that A is an $n \times n$ nonnegative matrix and that at least one of the conditions (2.6) or (2.7) holds. Then the matrix $(I - A)^{-1}$ exists and it is nonnegative.*

Proof. It follows from (2.5) that we have to show that $\lambda_A < 1$. We restrict ourselves only with the case (2.7). Let x_A be the eigenvector of A which corresponds to the eigenvalue λ_A, then it is nonnegative and

$$Ax_A = \lambda_A x_A.$$

Denote by $e = (1, \dots, 1)^T$ the unit vector in $\mathbb{R}^n$, then we have

$$e^T A x_A = \lambda_A \sum_{i=1}^{n} (x_A)_i.$$

Next, using (2.7), we obtain

$$e^T A x_A = \sum_{j=1}^{n} \left[\sum_{i=1}^{n} a_{ij} \right] (x_A)_j < \sum_{j=1}^{n} (x_A)_j. \tag{2.8}$$

Therefore, $\lambda_A < 1$, and the result follows now from Theorem 2.1. □

Exercise 2.2. Prove the assertion of Theorem 2.3 under condition (2.6).

Using a modification of the Perron-Frobenius theorem, we can somewhat strengthen the above result under additional assumptions on the economy model.

Definition 2.1. Let A be an $n \times n$ matrix. Then A is said to be *indecomposable*, if there is no index set $J \subsetneq \{1, \dots, n\}$ such that $a_{ij} = 0$ as $i \notin J, \quad j \in J$.

Thus, if A is an indecomposable matrix, it can not be rearranged into the form

$$\begin{pmatrix} A_1 & A_2 \\ 0 & A_3 \end{pmatrix},$$

where A_1 and A_3 are square submatrices. If A is also an input-output matrix, it means that there are no isolated production subsectors in the economy. In fact, otherwise for each $j \in J$, the j-th producer does not utilize any i-th commodity if $i \notin J$. In other words, if the input-output matrix is indecomposable, then all the commodities have either direct or indirect relations to each other. Now we recall the specialization of the Perron-Frobenius theorem for indecomposable matrices.

Theorem 2.4. *(Perron-Frobenius; e.g. see Nikaido (1968), Chapter 2. Theorems 7.3 and 7.4). Suppose A is an indecomposable nonnegative $n \times n$ matrix. Then:*

(i) it has a positive eigenvalue λ_A with a positive eigenvector, which is defined uniquely (up to scalar multiples), such that $\lambda_A \geq |\mu|$ for any eigenvalue of A;

(ii) the matrix $(I - A)^{-1}$ exists and positive if and only if $\lambda_A < 1$.

The indecomposability of A allows us to replace the conditions (2.6) and (2.7) with the following:

$$\sum_{j=1}^{n} a_{ij} \leq 1 \quad \text{for} \quad i = 1, \ldots, n \tag{2.6$'$}$$

and

$$\sum_{i=1}^{n} a_{ij} \leq 1 \quad \text{for} \quad j = 1, \ldots, n. \tag{2.7$'$}$$

More precisely, the sufficient condition for productivity is now formulated as follows.

Theorem 2.5. *Suppose that A is an indecomposable nonnegative $n \times n$ matrix and that either (2.6′) holds and at least one of the inequalities is strict or (2.7′) holds and at least one of the inequalities is strict. Then the matrix $(I - A)^{-1}$ exists and is positive.*

Proof. On account of Theorem 2.4 it suffices to prove that $\lambda_A < 1$. This fact is obtained along the lines of the proof of Theorem 2.3. Observe that x_A is now positive, hence (2.8) is also true. □

Exercise 2.3. Prove the inequality $\lambda_A < 1$ in Theorem 2.5.

Exercise 2.4. Prove that an $n \times n$ matrix A is indecomposable if, for each pair $(i, j), i \neq j$, there exist indices $i_1, i_2, \ldots, i_{s-1}$ such that $a_{i_{l-1} i_l} > 0$ for $l = 1, 2, \ldots, s$, where $i_0 = i$, $i_s = j$.

Thus, this connection between all the branches of an economy allows for satisfaction of any final demand if it is possible to create inventories for at least one commodity. Of course, this result strengthens essentially that in the general case.

2.2 Generalizations

So, system (2.1), which represents the classical model, can be reduced to the problem of solving the system of linear equations. This is mainly due to the fact that the matrix A has nonnegative entries (see (2.5)), hence the solution can be found from the explicit formula (2.3) which admits the detailed analysis of the problem, involving sensitivity issues. However, even small modifications of the model caused by the peculiarities of the real economic system require more general models and more powerful methods of their analysis, respectively. This does not mean that equilibrium disappears, but it has to be redefined in a new format.

For example, if we take into account environmental aspects, the nonnegativity of all input-output coefficients a_{ij} does not hold. Such a model was suggested by *W. Leontief and D. Ford.* In fact, a_{ij} determines the amount of the i-th commodity which is used for production of the j-th commodity. However, the production process results in by-products which are not used by other processes and require special utilization procedures since otherwise they pollute the environment. Nevertheless, these by-products should be included in the input-output model, but the coefficient a_{ij} then becomes nonpositive if the index i corresponds to the commodity which is a by-product in producing the j-th commodity. Due to (2.5), it means that the matrix $(I-A)^{-1}$ need not have only nonnegative entries, hence the model (2.1) can not be reduced to a system of linear equations in general. In this case the concept of a solution may be redefined in a more suitable format.

The next problem is possible limitations for outputs and resources. In fact, if we have certain values of outputs $\tilde{x}$ for the previous period, the outputs in the next period should be in a segment $[x', x'']$ containing $\tilde{x}$, since the technological processes do not admit large deviations. Then the system (2.1) (or (2.2)) should be replaced with the following:

$$(I-A)x = y, \quad x' \leq x \leq x''. \tag{2.2'}$$

Exercise 2.5. Write down the extension of the input-output model for the case (2.2′).

The usual resource constraints are of the form:

$$\sum_{j=1}^{n} c_{ij}x_j \leq d_i \quad i = 1, \dots, m; \tag{2.9}$$

where d_i is the total inventory of the i-th factor (resource) and c_{ij} is the stock of the i-th factor employed per unit of the j-th product. Usually, there are limitations on a capital stock and a labor force. The corresponding model can be formally written as the system (2.2), (2.9) or (2.2′), (2.9), but it can be inconsistent, hence the concept of a solution should be redefined.

For instance, one can consider the optimization problem

$$\begin{array}{ll} \max \rightarrow & \alpha \\ & \text{subject to} \\ & x - Ax \geq \alpha y, \\ & l^T x \leq L, \quad x \geq 0; \end{array}$$

where l is the vector of labor coefficients, L is the total labor force. This model maintains the proportions among commodities in final demand and maximizes the level of the final demand.

It is possible to consider also the problem of minimizing the total amount of some resources employed in production processes under the constraints of form (2.2′). Moreover, some technologies may be nonlinear, i.e. the model will involve a nonlinear map $A(x)$ instead of the linear map Ax of the production demands. Anyway, the modified models require new mathematical tools for their investigation and solution. These tools will be based on new concepts of solutions and will provide new conditions for existence and uniqueness results and algorithms for finding solutions.

2.3 Closed input-output model

The closed version of the input-output model corresponds to the case when the final demand is zero, i.e. the total output for each commodity is equal to its industrial demand. This model describes a closed system and is also called *the linear exchange model* or *the international trade model.* In the latter case, the input-output coefficient a_{ij} indicates the part of the total income of the j-th country for purchasing goods of the i-th country, i.e. it is supposed that the proportions between these parts are fixed from the previous time period. From the definition it follows that

$$\begin{array}{l} a_{ij} \geq 0 \quad \text{for} \quad i,j = 1,\dots,n \\ \qquad\qquad \text{and} \\ \sum\limits_{i=1}^{n} a_{ij} = 1 \quad \text{for} \quad i = 1,\dots,n; \end{array} \tag{2.10}$$

where n denotes the number of countries involved in the international trade system. Denote by π_i the total income of the i-th country, then

$$\pi_i \geq 0 \quad \text{and} \quad \pi_i \leq \sum_{j=1}^{n} a_{ij}\pi_j \quad \text{for} \quad i = 1,\dots,n. \tag{2.11}$$

It means that the income is due to the international trade only. Observe that (2.10) in fact implies the equalities in (2.11). Conversely, suppose that there exists an index s such that

$$\pi_s < \sum_{j=1}^{n} a_{sj}\pi_j,$$

then

$$\pi_i \leq \sum_{j=1}^{n} a_{ij}\pi_j \quad \text{for} \quad i \neq s.$$

Summing up all these inequalities, we obtain

$$\sum_{i=1}^{n} \pi_i < \sum_{i=1}^{n} \left(\sum_{j=1}^{n} a_{ij}\pi_j \right) = \sum_{j=1}^{n} \left(\sum_{i=1}^{n} a_{ij}\pi_j \right)$$
$$= \sum_{j=1}^{n} \pi_j \left(\sum_{i=1}^{n} a_{ij} \right) = \sum_{j=1}^{n} \pi_j$$

due to (2.10). However, this is a contradiction.

Thus, the problem is to find

$$\pi_i \geq 0 \quad \text{such that} \quad \pi_i - \sum_{j=1}^{n} a_{ij}\pi_j = 0 \quad \text{for } i = 1, \dots, m; \tag{2.12}$$

or equivalently,

$$(I - A)\pi = 0, \quad \pi \geq 0, \tag{2.12'}$$

where $\pi = (\pi_1, \dots, \pi_n)^T$ (cf. (2.1) and (2.2)). We have obtained a particular case of the open input-output model, but the additional property (2.10) of the matrix A enables us to provide a thorough analysis of its solution in a somewhat different way.

It is clear that system (2.12) (or (2.12′)) always has the trivial zero solution but it corresponds to the absence of any income for every country. Therefore, we have to answer the following two questions:

(Q1) *When system (2.12) has a nontrivial solution, i.e. there exist countries with positive incomes?*

(Q2) *When system (2.12) has a positive solution, i.e. all the countries have positive incomes?*

First we note that the matrix A has eigenvalue 1. In fact, set $e = (1, \dots, 1)^T \in \mathbb{R}^n$, then

$$A^T e = (e^T A)^T = e,$$

since

$$[e^T A]_i = \sum_{j=1}^{n} a_{ij} = 1 \quad \text{for} \quad i = 1, \dots, n$$

because of (2.10). It means that 1 is an eigenvalue of A^T, but the sets of eigenvalues for matrices A and A^T coincide. It follows that

$$|I - A| = 0,$$

i.e. the system

$$(I - A)\pi = 0 \tag{2.13}$$

has nontrivial solutions. We thus have to show that system (2.13) has the nontrivial solutions belonging to the set

$$\mathbb{R}^n_+ = \{x \in \mathbb{R}^n \,|\, x_i \geq 0 \quad i = 1, \ldots, n\}.$$

This assertion requires some additional results.

First we recall the well-known separability property of convex sets; see e.g. *Nikaido* (1968), Chapter 1, Theorem 3.2.

Proposition 2.1. *(Existence of a separating hyperplane) Suppose L is a nonempty, convex, and closed subset of $\mathbb{R}^n$. If $a \notin L$, then there exists an element $y \neq 0$ such that*

$$a^T y < l^T y \quad \textit{for every} \quad l \in L.$$

We utilize this property in the following lemma of the alternative.

Lemma 2.1. *Let B be a square $n \times n$ matrix. Then, either*
(i) $B^T x = 0$ for some $x \geq 0$, $\quad x \neq 0$;
or
(ii) $By > 0$ for some y.

Proof. Let b^i denote the i-th row of the matrix B. Then $b^i \in \mathbb{R}^n$ and the set

$$\begin{aligned} L &= \operatorname{conv}\left\{b^i\right\}_{i=1,\ldots,n} \\ &= \left\{ z \in \mathbb{R}^n \;\middle|\; z = \sum_{i=1}^n \mu_i b^i, \; \sum_{i=1}^n \mu_i = 1, \; \mu_i \geq 0 \; i = 1, \ldots, n \right\} \end{aligned}$$

is nonempty, convex, and closed. We consider two cases which are mutually exclusive: $0 \in L$ and $0 \notin L$. In the first case (see Figure 2.1) we have $0 = \sum_{i=1}^n \mu_i b^i$, $\sum_{i=1}^n \mu_i = 1$, $\mu_i \geq 0$ for $i = 1, \ldots, n$ and simply set $x = (\mu_1, \ldots, \mu_n)^T$, then (i) is true.

In the second case (see Figure 2.2), using Proposition 2.1, we obtain

$$l^T y > 0 \quad \text{for every} \quad l \in L,$$

i.e.

$$(b^i)^T y > 0 \quad \text{for} \quad i = 1, \ldots, n;$$

or equivalently,

$$By > 0$$

and (ii) is true. □

Figure 2.1: $0 \in L$

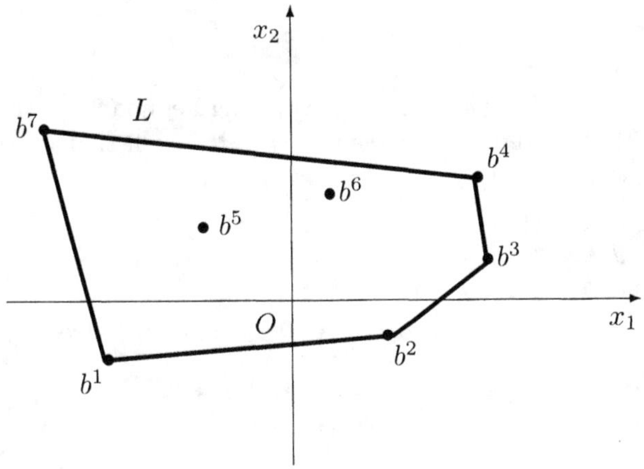

Figure 2.2: $0 \notin L$

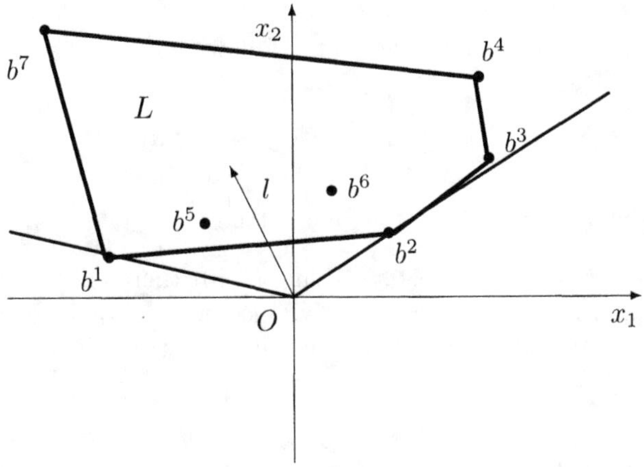

Exercise 2.6. Prove that the solution set of problem (2.12) is a convex cone.

We are now ready to give answers to both the questions.

Theorem 2.6. *If (2.10) holds, then system (2.12) always has a nontrivial solution.*

Proof. We argue by contradiction. Suppose that there is no such an element $\pi \geq 0, \quad \pi \neq 0$, that $(I-A)\pi = 0$. Applying Lemma 2.1 with $B = (I-A)^T$, we have that there exists an element $y \in \mathbb{R}^n$ such that $(I-A)^T y > 0$, or equivalently,

$$\sum_{i=1}^{n} y_i a^i < y \tag{2.14}$$

where a^i denotes the i-th row of A. Besides, (2.10) gives

$$\sum_{i=1}^{n} a^i = e = (1, \ldots, 1)^T. \tag{2.15}$$

Without loss of generality we suppose that y_1 is the smallest component of y, i.e.

$$\mu = y_1 = \min_{i=1,\ldots,n} y_i.$$

Multiplying (2.15) by μ and subtracting it from (2.14), we obtain the vector inequality

$$\sum_{i=2}^{n} (y_i - y_1) a^i < y - y_1 e.$$

The first scalar inequality here is the following

$$0 \leq \sum_{i=2}^{n} (y_i - y_1) a_{i1} < y_1 - y_1 = 0.$$

Hence, the assertion of the theorem is true. □

Thus, for each matrix A satisfying (2.10) there exists an international trade plan which guarantees positive incomes for some countries, i.e. the answer to (Q1) is positive. This result can be strengthened under an additional condition.

Theorem 2.7. *If the matrix A is indecomposable and satisfies (2.10), then system (2.12) has a positive solution which is defined uniquely up to scalar multiples.*

Proof. From Theorem 2.6 it follows that there exists an element $\pi \geq 0,\ \pi \neq 0$, such that $(I - A)\pi = 0$. Suppose that $\pi_j > 0$ for $j \in J$ and $\pi_j = 0$ for $j \in J'$ and $J' \neq \emptyset$. Then $J \cap J' = \emptyset, \quad J \cup J' = \{1, \ldots, n\}$ and (2.12) gives

$$\sum_{j=1}^{n} a_{ij}\pi_j = \sum_{j \in J} a_{ij}\pi_j = \pi_i = 0 \text{ for every } i \in J'.$$

Since $\pi_j > 0$ for $j \in J$, it follows that

$$a_{ij} = 0 \quad \text{for each pair} \quad (i, j) \in J' \times J.$$

Transposing rows and columns of the matrix A, we obtain then the matrix

$$\underbrace{\begin{pmatrix} A_1 & A_2 \\ 0 & A_3 \end{pmatrix}}_{J \quad J'} \begin{matrix} \}J \\ \}J' \end{matrix}$$

where A_1 and A_3 are square matrices. It means that A is decomposable, a contradiction, i.e. the first assertion of the theorem is true. Next, suppose that there exists a positive vector $\pi' \neq \pi$, which also satisfies (2.12). Set $\mu = \min_{i=1,\ldots,n} \pi_j/\pi'_j$ and, without loss of generality, suppose that $\mu = \pi_1/\pi'_1$. Then clearly $\mu > 0$ and

$$\pi''_j = \pi_j - \mu\pi'_j \begin{cases} = 0 & \text{if } \ j = 1, \\ \geq 0 & \text{if } \ j > 1. \end{cases}$$

Moreover, $A\pi'' = A\pi - \mu A\pi' = \pi - \mu\pi' = \pi''$, i.e. $\pi'' \neq 0$ also solves (2.12), but this contradicts the first part of the theorem since each nontrivial solution of (2.12) can not have zero components. Hence $\pi'' = 0$, i.e. $\pi = \mu\pi'$ and the proof is complete. □

Thus, if there are direct or indirect trade relationships between all the countries (no closed trade groups), then Theorem 2.7 says that all the countries receive positive incomes. It follows that even comparatively simple models can reveal deep relationships in the real system and support the world level decisions.

Chapter 3

Linear Dynamic Models of an Economy

In the previous chapter, we considered economic models for a fixed time period. However, we can also investigate the behavior of an economy for a comparatively long time, which corresponds to a model with the infinite number of time periods. For simplicity, we chose the models with discrete time. Again, we are interested in finding conditions which provide a stable or balanced work of the whole system. First we consider an extension of the open input-output model described in Section 2.1.

3.1 Extended dynamic input-output model

In the model, the economy is divided into n pure production sectors, each of them produces a homogeneous commodity. Recall that a_{ij} denotes the quantity of the i-th commodity for production of one unit of the j-th commodity, and these coefficients are supposed to be constant. It means that there are no significant changes in the production technologies. The static input-output model is then given in (2.1) (or (2.2)) and is rewritten as follows: For a given final demand $y = (y_1, \dots, y_n)^T$, find an output vector $x = (x_1, \dots, x_n)^T$ such that

$$(I - A)x = y, \quad x \geq 0, \tag{3.1}$$

where I is the $n \times n$ unit matrix, A is the $n \times n$ matrix with the entries a_{ij}. In Section 2.1, several sufficient conditions, which provide the existence of nontrivial solutions of this system for an arbitrary nonnegative value of the final demand, were presented. In the dynamic model, we consider the problem of existence of a satisfactory level of output which covers the industrial demand instead of (3.1). In other words, the problem is to find an output vector x such that

$$(I - A)x \geq 0, \quad x \geq 0. \tag{3.2}$$

Together with the natural balance in (3.2), we can consider its price counterpart, which can be formulated as follows: Find the price vector $p = (p_1, \dots, p_n)^T$ such that

$$(I - A)^T p \leq 0, \quad p \geq 0; \tag{3.3}$$

or equivalently

$$p_j \leq \sum_{i=1}^{n} a_{ij} p_i \text{ for } j = 1, \dots, n.$$

The j-th inequality means that the price of the j-th commodity does not exceed the cost of all the commodities which are used for production of one unit of the j-th commodity, i.e. there is no profit for each commodity.

Now we consider the infinite sequence of equal time periods so that x^t and p^t will denote the output and price vectors for the t-th time period, respectively. We suppose that the production in the $(t+1)$-th period uses only the commodities which were produced in the previous t-th period. Then the dynamic extension of the system (3.2) and (3.3) is the following

$$\left\{ \begin{array}{ll} x^t \geq A x^{t+1}, & x^t \geq 0, \\ p^{t+1} \leq A^T p^t, & p^t \geq 0, \end{array} \right. \text{ for } t = 0, 1, \dots; \tag{3.4}$$

where the starting values x^0 and p^0 are given. System (3.4) describes an economic system with non-decreasing outputs and non-increasing prices. Clearly, it always has the trivial solution, hence we are interested in finding nontrivial solutions. Observe that non-increasing prices and the absence of profits do not prevent the existence of such a system for an indefinitely long time while the input-output coefficients are constant.

For example, if one produces an output x' in the t-th period so that

$$c = (p^t)^T x' > (p^{t+1})^T x',$$

then it is possible to purchase commodities for producing any j-th commodity in the next (t+1)-th period such that

$$p_j^{t+1} = \sum_{i=1}^{n} a_{ij} p_i^t.$$

It means that the producer maintains the same capital value c in the next period, but he will be able to purchase a greater amount of commodities and thus support the extending level of production.

Thus, (3.4) describes a system which may be in an equilibrium state for a long time. This corresponds to the concept of a *dynamic equilibrium*. Moreover, we can investigate the specialized problem of existence of balanced trajectories $\{x^t\}$ and $\{p^t\}$ where the outputs and prices in neighboring periods differ only in scalar ratios. This problem will be solved for a somewhat more general economic model.

3.2 The von Neumann model of an expanding economy

In the mid-1930's, *John von Neumann* suggested a dynamic model of an economy, which extends the dynamic input-output model above. In describing this model, we follow *Lancaster* (1968) and *Nikaido* (1968).

The economy involves n commodities, but, unlike the usual input-output models, also m technologies of production and $m \neq n$ in general. The unit level of the j-th technology requires a_{ij} units of the i-th commodity and produces b_{ij} units of the i-th commodity, so that the system is determined with the help of the two $n \times m$ matrices A and B which represent pure industry consumption and production, respectively. For the infinite sequence of equal time periods, $x^t \in \mathbb{R}^m$ and $p^t \in \mathbb{R}^n$ will denote the activity levels and price vectors for the t-th period, respectively.

For clarity, we write analogues of conditions (3.4) in an expanded form:

$$\sum_{j=1}^{m} a_{ij}x_j^{t+1} \leq \sum_{j=1}^{m} b_{ij}x_j^t \; i = 1, \dots, n; \; x_j^t \geq 0 \quad j = 1, \dots, m; \tag{3.5}$$

$$\sum_{i=1}^{m} a_{ij}p_i^t \geq \sum_{i=1}^{m} b_{ij}p_i^{t+1} \; j = 1, \dots, m; \; p_i^t \geq 0 \quad i = 1, \dots, n; \tag{3.6}$$

for $t = 0, 1, \dots,$ with the known vectors x^0 and p^0. The first series of inequalities in (3.5) says that, for each i-th commodity, its industrial demand does not exceed its output in the previous time period. Also, the first series of inequalities in (3.6) says that the total cost of commodities produced per unit of each activity does not exceed the total cost of commodities utilized for one unit of this activity, i.e. the profits are absent for all technologies.

The system (3.5), (3.6) is expanded by including two conservation rules:

$$\sum_{i=1}^{n}\sum_{j=1}^{m} a_{ij}p_i^{t+1}x_j^{t+1} = \sum_{i=1}^{n}\sum_{j=1}^{m} b_{ij}p_i^{t+1}x_j^t \tag{3.7}$$

and

$$\sum_{j=1}^{m}\sum_{i=1}^{n} a_{ij}p_i^t x_j^{t+1} = \sum_{j=1}^{m}\sum_{i=1}^{n} b_{ij}p_i^{t+1} x_j^{t+1}. \tag{3.8}$$

Equality (3.7) says that the total cost of the industrial demand in the $(t+1)$-th period is equal to the total industrial income for commodities produced in the previous period, i.e. the industrial income is used only for satisfying the industrial demand. Equality (3.8) says that the industrial expenses in the $(t+1)$-th period (for the previous period prices) are equal to the total cost of outputs in this period, i.e. the total amount of money is constant.

We rewrite the system (3.5)–(3.8) in the equivalent form:

$$\begin{array}{l} Ax^{t+1} \leq Bx^t, (p^{t+1})^T Ax^{t+1} = (p^{t+1})^T Bx^t, x^t \geq 0; \\ A^T p^t \geq B^T p^{t+1}, (p^t)^T Ax^{t+1} = (p^{t+1})^T Bx^{t+1}, p^t \geq 0; \end{array} \tag{3.9}$$

for $t = 0, 1, \ldots$ Clearly, this system always has the trivial solution, hence we intend to find nontrivial solutions. Moreover, we will consider the balanced growth long-run trajectory which possesses the following properties:

$$x^{t+1} = \lambda x^t \text{ and } p^{t+1} = \mu^{-1} p^t, \tag{3.10}$$

where $\lambda > 0$ and $\mu > 0$. It means that all the activities have the same growth ratio $\lambda - 1$ and that all the prices decrease with the same ratio $\mu - 1$. Using (3.10) in (3.9) yields

$$\begin{array}{l} \lambda Ax \leq Bx, \ \lambda p^T Ax = p^T Bx, \ x \geq 0; \\ \mu A^T p \geq B^T p, \ \mu p^T Ax = p^T Bx, \ p \geq 0; \end{array} \tag{3.11}$$

where $x^t = \lambda^t x$ and $p^t = \mu^{-t} p$, and we also have to present conditions which provide existence of nontrivial solutions to system (3.11). Observe that (3.11) can be equivalently rewritten as the linear complementarity problem

$$z \geq 0, \ Qz \geq 0, \ z^T Qz = 0,$$

where $z = (x, p)$,

$$Q = \begin{pmatrix} 0 & (\mu A - B)^T \\ B - \lambda A & 0 \end{pmatrix},$$

and Q is skew-symmetric if $\lambda = \mu = \alpha$, i.e. $Q^T = -Q$.

John von Neumann was the first who gave the existence theorem for the above problem. Afterwards, the sufficient conditions in this theorem were somewhat relaxed by *J.G. Kemeny, O. Morgenstern, and G.L. Thompson.*

These conditions seem very natural and are the following.

(A1) *The matrices A and B contain only nonnegative entries.*

(A2) *The matrix A does not contain zero columns* (i.e. each activity utilizes a positive amount of at least one commodity).

(A3) *The matrix B does not contain zero rows* (i.e. production of each commodity requires a positive level of at least one technology).

The result is based on the following two lemmas.

Lemma 3.1. *Suppose C is an arbitrary $n \times m$ matrix. Then either*

(i) $p^T C \leq 0$ for some $p \geq 0$, $p \neq 0$,

or

(ii) $Cx > 0$ for some $x > 0$.

Proof. Let c^i denote the i-th row of the matrix C. By definition, $c^i \in \mathbb{R}^m$. Also, denote by e^i the i-th coordinate vector in $\mathbb{R}^m$, i.e.

$$e^i_j = \begin{cases} 0 & \text{if } i \neq j, \\ 1 & \text{if } i = j. \end{cases}$$

Set

$$L = \text{ conv}\{c^1, \ldots, c^n, e^1, \ldots, e^m\}.$$

By definition, the set L is nonempty, convex and closed. We consider two cases which are mutually exclusive: $0 \in L$ and $0 \notin L$. In the first case there exist numbers $\alpha_i \geq 0$ for $i = 1, \ldots, n$ and $\beta_j \geq 0$ for $j = 1, \ldots, m$ such that

$$\sum_{i=1}^{n} \alpha_i + \sum_{j=1}^{m} \beta_j = 1 \tag{3.12}$$

and

$$\sum_{i=1}^{n} \alpha_i c^i + \sum_{j=1}^{m} \beta_j e^j = 0,$$

or equivalently,

$$\sum_{i=1}^{n} \alpha_i c_{ij} + \beta_j = 0 \text{ for } j = 1, ..., m. \tag{3.13}$$

Set $p = (\alpha_1, \ldots, \alpha_n)^T \geq 0$. If $p = 0$, then (3.13) gives $\beta_j = 0$ for $j = 1, \ldots, n$, which contradicts (3.12). Therefore $p \neq 0$. From (3.13) we also obtain

$$\sum_{i=1}^{n} \alpha_i c_{ij} \leq 0 \text{ for } j = 1, \ldots, m,$$

or equivalently, $p^T C \leq 0$, i.e. (i) holds.

In the second case we use Proposition 2.1 and obtain

$$l^T y > 0 \text{ for all } l \in L.$$

Setting $x = y$ and using this inequality with $l = e^i$ gives $x_i > 0$. Then using the above inequality with $l = c^i$ gives

$$(c^i)^T x > 0 \text{ for } i = 1, \ldots, n,$$

i.e. $Cx > 0$ and (ii) holds. □

Clearly, Lemma 3.1 extends Lemma 2.1 from square matrices. Let us consider two optimization problems:

$$\max \;\rightarrow\; \lambda \tag{3.14}$$

subject to

$$\lambda Ax \leq Bx, \;\; x \geq 0, \;\; x \neq 0, \;\; \lambda > 0$$

and

$$\min \;\rightarrow\; \mu \tag{3.15}$$

subject to

$$\mu p^T A \geq p^T B, \;\; p \geq 0, \;\; p \neq 0, \;\; \mu > 0.$$

Lemma 3.2. *(i) Problems (3.14) and (3.15) always have solutions* (λ^*, x^*) *and* (μ^*, p^*).

(ii) It holds that $\lambda^* \geq \mu^*$.

Proof. First we replace (3.14) and (3.15) with the equivalent ones:

$$\max \;\rightarrow\; \lambda \tag{3.14$'$}$$

subject to

$$\lambda Ax \leq Bx, \;\; x \geq 0, \;\; e^T x = 1, \;\; \lambda > 0,$$

and

$$\min \;\rightarrow\; \mu \tag{3.15$'$}$$

subject to

$$\mu p^T A \geq p^T B, \;\; p \geq 0, \;\; e^T p = 1, \;\; \mu > 0;$$

where e is the corresponding vector of units. That is, if $(\bar{\lambda}, \bar{x})$ and $(\bar{\mu}, \bar{p})$ solve (3.14′) and (3.15′), they also solve (3.14) and (3.15). Conversely, if (λ^*, x^*) and (μ^*, p^*) solve (3.14) and (3.15), then $(\lambda^*, \bar{x})$ and $(\mu^*, \bar{p})$ with $\bar{x} = \frac{1}{e^T x^*} x^*$ and $\bar{p} = \frac{1}{e^T p^*} p^*$ solve (3.14′) and (3.15′), respectively.

Next, due to (A3), B contains only nonzero rows, hence there is a vector $x \geq 0$, $x \neq 0$ such that $Bx > 0$. Choosing λ small enough, we obtain $\lambda Ax \leq Bx$, i.e. the feasible sets in (3.14) and (3.14′) are nonempty. Take an arbitrary point $x \in \mathbb{R}^m$ such that $x \geq 0$ and $e^T x = 1$. Denote by l the number of its maximal component, i.e. $x_l = \max_{i=1,\dots,m} x_i$. Since A contains only nonzero columns, each column of the matrix $(B - \lambda A)$ will contain a negative element which is greater than the sum of the positive entries of the same row, if we choose λ large enough. Hence, there exists a number i such that the negative element has the numbers (i, l). It follows that

$$\sum_{j=1}^{m} (b_{ij} - \lambda a_{ij}) x_j < 0,$$

and the variable λ is bounded from above. Therefore, problem (3.14′) as well as (3.14) is solvable.

Noticing that (3.15) is equivalent to the problem:

$$\max \rightarrow \tau$$

subject to

$$p^T A \geq \tau p^T B, \; p \geq 0, \; p \neq 0, \; \tau > 0,$$

we obtain the existence of a solution to (3.15) along the same lines by using (A2) instead of (A3). So, assertion (i) is true.

Let us consider the system

$$(B - \lambda^* A)x > 0.$$

If it has a nonnegative solution $\tilde{x}$, then $\tilde{x} \neq 0$ and

$$(B - \lambda' A)\tilde{x} \geq 0$$

for some $\lambda' > \lambda^*$, which contradicts the definition of λ^*. Using now Lemma 3.1 with $C = B - \lambda^* A$, we conclude that there exists a vector $p \geq 0$, $p \neq 0$ such that

$$p^T (B - \lambda^* A) \leq 0,$$

i.e. the pair (λ^*, p) is feasible in (3.15). From the definition of μ^* we obtain $\mu^* \leq \lambda^*$, i.e. assertion (ii) is also true. □

We are now ready to establish the basic existence and uniqueness results for system (3.11).

Theorem 3.1. *There exist a number $\alpha > 0$ and vectors $x \in \mathbb{R}^m$, $p \in \mathbb{R}^n$ such that*

$$\begin{array}{l} x \geq 0,\ x \neq 0,\ \alpha Ax \leq Bx,\ p^T(\alpha A - B)x = 0, \\ p \geq 0,\ p \neq 0,\ \alpha p^T A \geq p^T B. \end{array} \tag{3.16}$$

Proof. By Lemma 3.2, there exist solutions (λ^*, x^*) and (μ^*, p^*) of problems (3.14) and (3.15), respectively, such that $\lambda^* \geq \mu^*$. Take any $\alpha \in [\mu^*, \lambda^*]$, then $x^* \geq 0$, $x^* \neq 0$, $p^* \geq 0$, $p^* \neq 0$ and we have

$$\alpha Ax^* \leq \lambda^* Ax^* \leq Bx^* \text{ and } \alpha(p^*)^T A \geq \mu^*(p^*)^T A \geq (p^*)^T B.$$

It follows that

$$(\alpha A - B)x^* \leq 0 \text{ and } (p^*)^T(\alpha A - B) \geq 0.$$

Multiplying these inequalities by $(p^*)^T$ and x^*, respectively, we obtain

$$0 \geq (p^*)^T(\alpha A - B)x^* \geq 0,$$

and the result follows. □

Thus, within assumptions (A1)–(A3), the economic system has a stationary long-run path which can be regarded as an example of a dynamic equilibrium. Moreover, this path admits the same ratio both for the balanced growth of production and for the balanced decrease of all prices. Recall that the vectors x and p in (3.11) (or (3.16)) represent the starting activity levels and prices.

We can somewhat strengthen the above result under the additional indecomposability assumption.

Definition 3.1. The set of matrices (A, B) is said to be *indecomposable*, if there are no nonempty index subsets $K \subseteq \{1, ..., n\}$ and $J \subseteq \{1, ..., m\}$ such that

$$\sum_{j \in J} a_{ij} = \sum_{j \in J} b_{ij} = 0 \text{ for each } i \notin K.$$

The indecomposability of (A, B) means that there is not a nonempty subset of commodity indices such that they can be produced without at least one commodity from the complement of this subset. In other words, there are either direct or indirect relations between all the commodities involved in this economy. This property enables us to establish the uniqueness of the ratio α for each starting pair (x, p).

Theorem 3.2. *If the set of the matrices (A, B) is indecomposable, then there exist $x \in \mathbb{R}^m$, $p \in \mathbb{R}^n$ and $\alpha > 0$ satisfying (3.16) with α being uniquely defined for each pair (x, p).*

Proof. The existence of x, p, α satisfying (3.16) has been proved in Theorem 3.1. By definition, $x \geq 0$, $x \neq 0$, hence $(b^i)^T x \geq 0$ for $i = 1, \dots, n$, where b^i denotes the i-th row of the matrix B. Since B contains only nonzero rows, there exists an index l such that $(b^l)^T x > 0$. Suppose there exists an index k such that $(b^k)^T x = 0$, then, using (3.16), we have

$$0 = (b^k)^T x \geq \alpha (a^k)^T x \geq 0,$$

i.e. $(a^k)^T x = 0$. It follows that the set of matrices (A, B) is decomposable with $K = \{1, \dots, n\} \backslash \{k\}$ since the k-th commodity does not participate in this system of technologies. By contradiction, we thus obtain $(b^i)^T x > 0$ for each $i = 1, \dots, n$, hence

$$p^T Bx > 0 \text{ and } p^T Ax > 0$$

since $\alpha > 0$. Thus, $\alpha = (p^T Bx)/(p^T Ax)$ is unique. □

Observe that the vectors x and p in (3.16) may be multiplied by any positive scalars without any changes for α. Under certain additional assumptions, the inequality $\alpha > 1$ may be also established.

Let us now consider the dynamic input-output model. Of course, we can also add the conservation rules to inequalities (3.4), then the problem becomes precisely the particular case of (3.9) where $B = I$, A is an $n \times n$ matrix. The stationary problem (3.11) can be then rewritten as follows: Find $x, p \in \mathbb{R}^n$ and $\lambda, \mu > 0$ such that

$$\begin{array}{l} \lambda Ax \leq x, \ \lambda p^T Ax = p^T x, \ x \geq 0, \ x \neq 0; \\ \mu A^T p \geq p, \ \mu p^T Ax = p^T x, \ p \geq 0, \ p \neq 0. \end{array}$$

If we suppose that A has only nonnegative entries and does not contain zero columns, then conditions (A1)–(A3) will be satisfied.

Exercise 3.1. Prove that the indecomposability of the set (A, I) is equivalent to the indecomposability of the non-negative matrix A.

Hence, Theorem 3.1 then states that, under the general assumptions, there exist a number $\alpha > 0$ and vectors $x, p \in \mathbb{R}^n$ such that

$$\begin{array}{l} x \geq 0, \ x \neq 0, \ \alpha Ax \leq x, \ p^T(\alpha A - I)x = 0, \\ p \geq 0, \ p \neq 0, \ \alpha p^T A \geq p^T. \end{array}$$

Due to Theorem 3.2, the number α in this system is uniquely defined for each pair x, p, if A is indecomposable.

Thus, the results above give a basis for indicating conditions of existence of stable equilibria for dynamic systems. Their extensions can be found e.g in the books by *Morishima* (1964) and *Nikaido* (1968).

Chapter 4

Optimization and Equilibria

Linear programming is known as one of the most effective tools for formulation and solution of many optimization type problems arising in various fields of applications; see e.g. *Dantzig* (1963) and *Schrijver* (1986) and references therein. We are interested in considering economic applications of linear programming which are regarded as a bridge to more general economic equilibrium problems.

4.1 Linear programming problems

We begin our considerations from the classical problem of income maximization subject to resource limitations. In this model we are given a firm that may produce n commodities for a fixed time period and utilize m factors (resources). Next, c_j denotes the price of the j-th commodity, b_i denotes the endowment of the i-th resource, and a_{ij} denotes the amount of the i-th resource for producing one unit of the j-th commodity, and all these values are supposed to be fixed for the time period. Then the problem above can be written as follows:

$$\max \rightarrow \sum_{j=1}^{n} c_j x_j \tag{4.1}$$

subject to

$$\begin{aligned} \sum_{j=1}^{n} a_{ij} x_j \leq b_i \quad & \text{for } i = 1, \ldots, m; \\ x_j \geq 0 \quad & \text{for } j = 1, \ldots, n; \end{aligned} \tag{4.2}$$

where x_j denotes the unknown output of the j-th commodity. Problem (4.1), (4.2) is nothing but the classical *linear programming problem* in the standard form. We can define the vectors $c = (c_1, \ldots, c_n)^T$, $x = (x_1, \ldots, x_n)^T$ and $b = (b_1, \ldots, b_m)^T$ and the $m \times n$ matrix A with the entries a_{ij}, and rewrite (4.1), (4.2) briefly:

$$\max \rightarrow c^T x \tag{4.1'}$$

subject to

$$Ax \leq b, x \geq 0. \tag{4.2'}$$

By definition, the utility function in (4.1) gives the total cost of commodities, whereas the first series of inequalities in (4.2) shows that the industrial demand of each factor can not exceed its endowment, and the second series simply determines the firm as producer of commodities.

In order to formulate optimality conditions for linear programming problems, one can utilize results of *the duality theory.* For instance, the dual problem of (4.1),(4.2) is written as follows

$$\min \rightarrow \sum_{i=1}^{m} b_i y_i \tag{4.3}$$

subject to

$$\begin{aligned} &\sum_{i=1}^{m} a_{ij} y_i \geq c_j \quad \text{for } j = 1, \ldots, n; \\ &\quad y_i \geq 0 \quad \text{for } i = 1, \ldots, m. \end{aligned} \tag{4.4}$$

Using the notation $y = (y_1, \ldots, y_m)^T$ we also can rewrite the above problem briefly:

$$\min \rightarrow b^T y \tag{4.3'}$$

subject to

$$A^T y \geq c, y \geq 0. \tag{4.4'}$$

The rules for creating a dual problem are clear: we replace the coefficients in the cost function and the right-hand sides of constraints, the number of dual variables corresponds to the number of constraints in the primal problem, we use the opposite operation for the cost function and the opposite inequality signs in constraints. Following these formal rules, we notice that the dual problem of (4.3),(4.4) is equivalent to (4.1),(4.2), hence both the problems can be termed as a primal-dual pair of linear programming problems. In what follows, we will use the following notation:

$$D = \{x \in \mathbb{R}^n \mid Ax \leq b, x \geq 0\}$$

and

$$\widetilde{D} = \{y \in \mathbb{R}^m \mid A^T y \geq c, y \geq 0\}$$

for the feasible sets of the primal and dual problems. We denote by D^* and $\widetilde{D}^*$ the solutions sets of these problems, respectively. Also, we denote by a^i and A^j the i-th row and the j-th column of the matrix A, respectively.

Lemma 4.1. *For arbitrary points $x \in D$ and $y \in \widetilde{D}$, it holds that*

$$c^T x \leq b^T y. \tag{4.5}$$

Proof. Fix $x \in D$ and $y \in \widetilde{D}$. Then, by definition,

$$\begin{aligned} c^T x &= \sum_{j=1}^{n} c_j x_j \leq \sum_{j=1}^{n} \left(\sum_{i=1}^{m} a_{ij} y_i \right) x_j \\ &= \sum_{i=1}^{m} \left(\sum_{j=1}^{n} a_{ij} x_j \right) y_i \leq \sum_{i=1}^{m} b_i y_i = b^T y, \end{aligned} \tag{4.6}$$

i.e. (4.5) holds. □

This simple lemma yields several very useful properties, including a sufficient condition of optimality.

Theorem 4.1. *If $x^* \in D$ and $y^* \in \widetilde{D}$ and*

$$c^T x^* = b^T y^*, \tag{4.7}$$

then $x^* \in D^*$ and $y^* \in \widetilde{D}^*$.

The proof follows directly from (4.7) and (4.5).

Theorem 4.2. *(i) If the cost function in (4.1) is not bounded from above on the feasible set, then $\widetilde{D} = \emptyset$.*

(ii) If the cost function in (4.3) is not bounded from below on the feasible set, then $D = \emptyset$.

The proof can be obtained by contradiction from Lemma 4.1.

Exercise 4.1. Prove the assertions of Theorems 4.1 and 4.2.

The derivation of necessary conditions of optimality is not so easy and is usually based on a suitable version of the *Farkas* Lemma. We follow the approach from *Eremin* (1998).

Proposition 4.1. *(See Eremin (1998), Theorem 5.2 and also Schrijver (1986), Corollary 7.1 H). let the system $\widetilde{A}x \leq \widetilde{b}$ be consistent for an $l \times n$ matrix $\widetilde{A}$, and a vector $\widetilde{b} \in \mathbb{R}^l$. Then, for a number $\alpha > 0$, the implication*

$$\widetilde{A}x \leq \widetilde{b} \Longrightarrow c^T x \leq \alpha$$

holds if and only if

$$c^T x - \alpha \equiv \sum_{i=1}^{l} y_i \left((\widetilde{a}^i)^T x - \widetilde{b}_i \right) - y_0$$

for some $y_i \geq 0$, $i = 0, \ldots, l$.

We are now ready to establish the basic theorem of duality.

Theorem 4.3. *If* x^* *solves the problem (4.1),(4.2), then there exists a solution* y^* *of the dual problem (4.3),(4.4), so that (4.7) holds.*

Proof. Applying Proposition 4.1 with $\alpha = c^T x^*$,

$$\widetilde{A} = \begin{pmatrix} A \\ -I \end{pmatrix} \quad \text{and} \quad \widetilde{b} = \begin{pmatrix} b \\ 0 \end{pmatrix},$$

we see that

$$c^T x - \alpha \equiv \sum_{i=1}^{m} y_i^* \left((a^i)^T x - b_i \right) - \sum_{j=1}^{n} v_j^* x_j - y_0^*$$

where $y_i^* \geq 0$ for $i = 0, 1, \ldots, m$ and $v_j^* \geq 0$ for $j = 1, \ldots, n$. If we suppose that $y_0^* > 0$, it follows that

$$x \geq 0, Ax \leq b \Longrightarrow c^T x \leq c^T x^* - y_0^* < c^T x^*$$

for every x, which is a contradiction. Therefore, $y_0^* = 0$ and we have

$$c - A^T y^* = -v^* \leq 0 \text{ and } c^T x^* = b^T y^*.$$

Thus, $y^* \in \widetilde{D}$ and Theorem 4.1 now gives $y^* \in \widetilde{D}^*$. □

The optimality condition (4.7) in Theorems 4.1 and 4.3 can be replaced with the so-called complementarity slackness conditions:

$$\begin{array}{ll} x_j^* \left[(A^j)^T y^* - c_j \right] = 0 & \text{for } j = 1, \ldots, n; \\ y_i^* \left[(a^i)^T x^* - b_i \right] = 0 & \text{for } i = 1, \ldots, m. \end{array} \tag{4.8}$$

In fact, if $x^* \in D$, $y^* \in \widetilde{D}$ and (4.8) holds, then all the inequalities in (4.6) with $x = x^*$ and $y = y^*$ are transformed into equalities and we obtain (4.7). Conversely, if $x^* \in D$, $y^* \in \widetilde{D}$ and (4.7) holds, then this implies the equalities in (4.6) with $y = y^*$, $x = x^*$ and we obtain (4.8). For simplicity, we shall formulate the corresponding result in a somewhat weakened, but symmetric form.

Theorem 4.4. *Points $x^* \in \mathbb{R}^n$ and $y^* \in \mathbb{R}^m$ solve problems (4.1),(4.2) and (4.3),(4.4), respectively, if and only if*

$$\begin{array}{lll} x^* \geq 0, & Ax^* \leq b, & (y^*)^T[Ax^* - b] = 0; \\ y^* \geq 0, & A^T y^* \geq c, & (x^*)^T[A^T y^* - c] = 0. \end{array} \tag{4.9}$$

Similarly to the stationary growth problem, this system can be equivalently rewritten as the linear complementarity problem:

$$z^* \geq 0, Qz^* - q \geq 0, (z^*)^T(Qz^* - q) = 0; \tag{4.10}$$

where $z^* = (x^*, y^*) \in \mathbb{R}^{n+m}$,

$$q = \begin{pmatrix} c \\ -b \end{pmatrix}, \quad Q = \begin{pmatrix} 0 & A^T \\ -A & 0 \end{pmatrix}.$$

Observe that the matrix Q is skew-symmetric.

4.2 Economic interpretation of optimality conditions

The optimality conditions given in Theorems 4.1 – 4.3 allow us to explain the economic sense of all the components of the dual problem. In fact, condition (4.7) shows that the utility function in the dual problem (4.3),(4.4) also indicates the cost value. Moreover, if we consider endowments of factors as variables, then

$$f^*(b) = c^T x^* = b^T y^*$$

and

$$\frac{\partial f^*(b)}{\partial b_i} = y_i^*,$$

i.e. y_i^* determines the sensitivity of the maximal income with respect to small changes of the i-th factor. It means that y_i can be defined as the cost (shadow price) of one unit of the i-th factor. Then, each j-th constraint in (4.4) means that the price of the j-th commodity cannot exceed the total cost of factors utilized for producing one unit of this commodity. Thus, the dual problem (4.3),(4.4) consists in minimizing the total cost of resources under the above constraints, including also the nonnegativity conditions for factor costs.

Using these definitions as a basis, we are now in a position to give an economic interpretation of optimality conditions (4.9). Of course, first two columns in (4.9) give the usual feasibility conditions of both the linear programming problems. The third column in (4.9), which is now equivalent to (4.8), gives just the complementarity slackness conditions of optimality for feasible points. Namely, if the j-th commodity is included in the optimal output ($x_j^* > 0$), then the price of this commodity is equal to the total cost of factors per unit of this commodity, and, conversely, if the price is less than the total cost of factors per unit, then the j-th commodity is not included in the optimal output. Next, if the shadow price of i-th factor is positive, then the precise balance holds between the industry demand and the endowment of this factor, and, conversely, if the industry demand is less than the endowment, then the shadow price of the factor is zero, this corresponds to the incomplete utilization of the i-th factor.

Moreover, the set of optimality conditions (4.9) can be regarded as a separate equilibrium model. Developing this approach, let us introduce an auxiliary vector $p = (p_1, \dots, p_m)^T$ of factors' prices. Next, suppose that the vectors c and b are not fixed in general, but commodity prices are in general dependent of outputs, and resource endowments are in general dependent of prices of factors. In other words, it is assumed that there exist mappings $c : \mathbb{R}^n_+ \to \mathbb{R}^n$ and $b : \mathbb{R}^m_+ \to \mathbb{R}^m$. The equilibrium conditions can be then rewritten as follows:

$$\begin{array}{l} x^* \geq 0, Ax^* \leq b(p^*), (y^*)^T[Ax^* - b(p^*)] = 0; \\ y^* \geq 0, A^T y^* \geq c(x^*), (x^*)^T[A^T y^* - c(x^*)] = 0; \end{array}$$

with addition of the coincidence rule between shadow and market prices of factors: $p^* = y^*$. This is nothing but the direct extension of conditions (4.9). Thus, we can apply this approach for much more general problems. It is suitable to determine the equilibrium conditions only for the pair (x^*, y^*):

$$\begin{array}{ll} x^* \geq 0, & y^* \geq 0; \\ A^T y^* - c(x^*) \geq 0, & b(y^*) - Ax^* \geq 0; \\ (x^*)^T[A^T y^* - c(x^*)] = 0, & (y^*)^T[b(y^*) - Ax^*] = 0. \end{array} \tag{4.11}$$

This system can be equivalently rewritten as the nonlinear complementarity problem:

$$z^* \geq 0, F(z^*) \geq 0, (z^*)^T F(z^*) = 0; \tag{4.12}$$

cf. (4.10), where $z = (x, y) \in \mathbb{R}^{n+m}$, $F(z) = Qz - q(z)$, and

$$q(z) = \begin{pmatrix} c(x) \\ -b(y) \end{pmatrix}, \quad Q = \begin{pmatrix} 0 & A^T \\ -A & 0 \end{pmatrix}.$$

If a pair (x^*, y^*) solves system (4.11), then it solves system (4.9) where $b = b(y^*)$ and $c = c(x^*)$. This means that x^* and y^* are solutions of problems (4.1′), (4.2′) and (4.3′), (4.4′), respectively, with $b = b(y^*)$ and $c = c(x^*)$, which may be treated as implicit optimization problems since the coefficients in the cost functions and the right-hand side of constraints depend on the unknown solutions. We see that problem (4.11) represents a model of the system whose behavior and structure may change, i.e. they depend on certain external parameters. Therefore, unlike the previous "pure" optimization models (4.1′), (4.2′) and (4.3′), (4.4′) with fixed parameters c and b, one must take into account the reaction of the system, and the optimal solution now corresponds to a kind of an equilibrium state of the system. Many general equilibrium problems, which will be described in the next chapters, reduce to the format of system (4.11). It follows that the theory and methods of complementarity problems may serve as a basis for investigation of these equilibrium models.

4.3 Economic interpretation of the solution method

Being based on the optimality conditions, we can construct an iterative solution method for solving problems (4.1),(4.2) or (4.3),(4.4). Let us consider a simple numerical example with linear programming problems.

Example 4.1. Set $m = n = 2$, $c = (2, 3)^T$, $b = (7, 18)^T$,

$$A = \begin{pmatrix} 1 & 2 \\ 4 & 3 \end{pmatrix},$$

i.e. the primal problem is the following:

$$\max \rightarrow 2x_1 + 3x_2$$

subject to

$$\begin{aligned} &x_1 + 2x_2 \leq 7, \\ &4x_1 + 3x_2 \leq 18, \\ &x_1 \geq 0, x_2 \geq 0. \end{aligned}$$

This problem can be solved by the graphical method (see Figure 4.1).

The feasible set is represented by the polygon OABC. The maximal value of the cost function is attained at the extremal point B $(3, 2)$ and results in 12. Let us consider the dual problem:

$$\min \rightarrow 7y_1 + 18y_2$$

subject to

$$\begin{aligned} &y_1 + 4y_2 \geq 2, \\ &2y_1 + 3y_2 \geq 3, \\ &y_1 \geq 0, y_2 \geq 0. \end{aligned}$$

Figure 4.1:

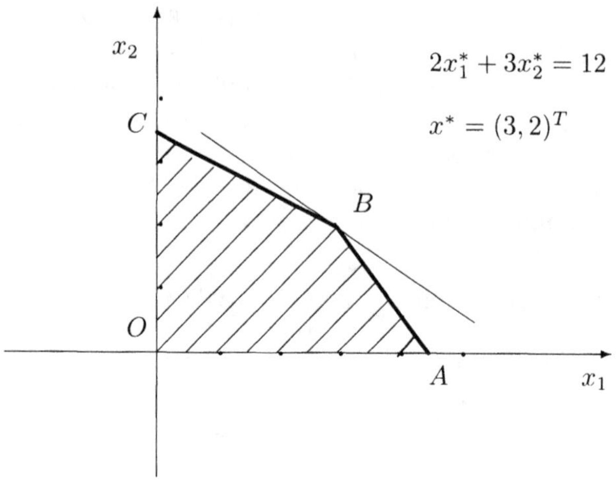

Since the optimal outputs x_1^*, x_2^* are positive, (4.8) yields

$$\begin{cases} y_1^* + 4y_2^* = 2, \\ 2y_1^* + 3y_2^* = 3. \end{cases}$$

This system has the unique solution $y^* = (1.2, 0.2)^T$. Since y^* is nonnegative it has to be a solution to the dual problem. In fact,

$$7y_1^* + 18y_2^* = 12$$

and (4.7) holds.

Exercise 4.2. Find a solution to the following linear programming problem:

$$\max \;\rightarrow\; 2x_1 + x_2$$

subject to

$$\begin{aligned} &3x_1 + 7x_2 \leq 21, \\ &6x_1 + 3x_2 \leq 18, \\ &8x_1 + 7x_2 \leq 28, \\ &x_1 \geq 0, x_2 \geq 0; \end{aligned}$$

and a solution to its dual.

This simple example suggests an idea of a solution method for the linear programming problem in the general case. In the two-dimensional case, the feasible set is always polygonal, hence if the problem is solvable, then solutions are common points of the feasible set and the corresponding tangent level line. Obviously, these solutions must contain an extreme point. In the general case, the feasible set D is polyhedral and each level set

$$\{x \mid c^T x = \alpha\}$$

is a hyperplane, therefore, if the solution set D^* is nonempty, it also must contain an extreme point of D. The number of such extreme points is always finite, hence we have to examine only the extreme points of the feasible set with ascent of the cost function if possible and we obtain a finite solution method for general linear programming problems. This is the key idea of the famous *simplex method* (see *Dantzig* (1963) for details). Using the example, we can even suggest a way of generating extreme points. In fact, we can rewrite the primal problem in the so-called canonical form

$$\max \;\to\; 2x_1 + 3x_2$$

subject to

$$\begin{aligned} & x_1 + 2x_2 + u_1 = 7, \\ & 4x_1 + 3x_2 + u_2 = 18, \\ & x_1 \geq 0, x_2 \geq 0, u_1 \geq 0, u_2 \geq 0; \end{aligned}$$

by introducing the auxiliary variables u_1 and u_2. The coefficient matrix

$$A = \begin{pmatrix} 1 & 2 & 1 & 0 \\ 4 & 3 & 0 & 1 \end{pmatrix}$$

of this problem has rank 2. If we choose two independent columns of this matrix as the basic ones, then setting the other (non-basic) variables to be zero yields a (unique) solution of the system of linear equations corresponding to some extreme point. In general, we can include the auxiliary variables x_{n+i}, $i = 1, \ldots, m$ in (4.2), thus obtaining the problem:

$$\max \;\to\; \sum_{j=1}^{n} c_j x_j$$

subject to

$$\begin{aligned} & \sum_{j=1}^{n} a_{ij} x_j + x_{n+i} = b_i \quad \text{for } i = 1, \ldots, m; \\ & x_j \geq 0 \text{ for } \; j = 1, \ldots, n+m; \end{aligned}$$

then the $m \times (n+m)$ matrix $\widetilde{A} = (A \mid I)$ has rank m. We call the index set B *basic* if it contains indices of m linearly independent columns of $\widetilde{A}$. A basic set B is called feasible if the system

$$\sum_{j \in B} x_j \widetilde{A}^j = b \tag{4.13}$$

has a nonnegative solution.

We now give an economic description of the simplex method. Each iteration begins from a feasible basic set B, i.e. we have a basic collection of m commodities, whose production demands do not exceed the endowment, i.e. (4.13) holds. We set $x_j = 0$ for $j \notin B$ and find shadow prices of factors for the current basic collection of commodities from the system

$$(\widetilde{A^j})^T y = c_j \quad \text{for } j \in B,$$

which also has a unique solution. For this pair (x, y) we have

$$c^T x = \sum_{j \in B} c_j x_j = \sum_{j \in B} x_j (\widetilde{A^j})^T y = b^T y,$$

hence if the dual vector satisfies the dual constraints which are now the following:

$$\widetilde{A^T} y \geq c,$$

but in fact it suffices to verify only the part of inequalities:

$$(\widetilde{A^j})^T y \geq c_j \quad \text{for } j \notin B,$$

then x is a solution to the initial problem. It means that the total costs of factors per unit of all nonbasic commodities with utilizing current shadow prices are not less then the prices of these commodities. Otherwise, there exists at least one commodity $k \notin B$, such that $(\widetilde{A^k})^T y < c_k$, then we can increase the total income by including this commodity and re-distributing the resources. We determine the subset

$$B^+ = \left\{ i \in B \;\middle|\; g_{ik} > 0, \text{ where } \sum_{j \in B} g_{ik} \widetilde{A^i} = \widetilde{A^k} \right\}.$$

This subset indicates the basic commodities which can be in principle replaced with the k-th commodity with nondecreasing total income. Note that the situation $B^+ = \emptyset$ means that the cost function is not bounded from above. We choose the index $l \in B^+$ with the property:

$$\theta = \frac{x_l}{g_{lk}} = \min_{i \in B^+} \frac{x_i}{g_{ik}},$$

whose removing yields the maximal profit, then we obtain the new feasible basic set $B' = B\backslash\{l\}\bigcup\{k\}$ and the new output

$$x'_i = \begin{cases} x_i - \theta g_{ik} & \text{if } i \in B, \\ \theta & \text{if } i = k, \\ 0 & \text{if } i \notin B \text{ and } i \neq k. \end{cases}$$

It follows that

$$c^T x' = c^T x + \theta[c_k - \widetilde{A^k y}] > c^T x,$$

i.e. the total income increases at the new extreme point, which corresponds to the new basic collection of commodities.

From the above result we also deduce that despite the possible value of n, we can always provide the maximal income by producing only basic commodities. It follows that there exist rather simple procedures for solving affine problems and these procedures do not require additional assumptions for finding a solution. However, if the affine problems do not represent a satisfactory approximation of the real system, they are replaced with more complicated nonlinear models which need new methods of investigation and solution.

Chapter 5

Nonlinear Economic Equilibrium Models

In this chapter, we consider several classes of nonlinear economic equilibrium models, which can be regarded as some extensions and modifications of the equilibrium conditions for linear programming problems.

5.1 Cassel-Wald type economic equilibrium models

We first consider a class of economic equilibrium models, which can be viewed as a modification of the known *Cassel-Wald* model.

The model describes an economic system which deals in n commodities and m pure factors of production. In what follows, c_k denotes the price of the k-th commodity, b_i denotes the total inventory of the i-th factor, and a_{ij} denotes the consumption rate of the i-th factor which is required for producing one unit of the j-th commodity, so that we set $c = (c_1, \dots, c_n)^T$, $b = (b_1, \dots, b_m)^T$, $A = (a_{ij})_{m \times n}$. Next, x_j denotes the output of the j-th commodity and p_i denotes the (shadow) price of the i-th factor, so that $x = (x_1, \dots, x_n)^T$ and $p = (p_1, \dots, p_m)^T$. The vector b is fixed, but this is not the case for c, i.e. it is assumed that there exists a mapping $c : \mathbb{R}^n_+ \to \mathbb{R}^n_+$. This means that prices are dependent of outputs.

The pair (x^*, p^*) is said to be in equilibrium if the following relations hold:

$$\begin{array}{ll} x^* \geq 0, & p^* \geq 0; \\ A^T p^* - c(x^*) \geq 0, & b - Ax^* \geq 0; \\ (x^*)^T [A^T p^* - c(x^*)] = 0, & (p^*)^T [b - Ax^*] = 0. \end{array} \tag{5.1}$$

This system is a particular case of the equilibrium problem (4.11). Obviously, in the case where the vector c is also fixed, system (5.1) becomes equivalent to the following pair of linear programming problems:

$$\begin{array}{ccc} \max \to c^T x & \text{and} & \min \to b^T p \\ \text{subject to} & & \text{subject to} \\ Ax \leq b, x \geq 0; & & A^T p \geq c, p \geq 0. \end{array}$$

However, in the general case, we can rewrite (5.1) equivalently as follows:

$$\begin{array}{l} (x^* - x)^T c(x^*) + (Ax - Ax^*)^T p^* \geq 0 \quad \forall x \geq 0; \\ (p - p^*)^T (b - Ax^*) \geq 0 \quad \forall p \geq 0; \end{array}$$

this is nothing but the optimality condition for the variational inequality problem: Find $x^* \in D$ such that

$$(x^* - x)^T c(x^*) \geq 0 \quad \forall x \in D, \tag{5.2}$$

where

$$D = \{x \in \mathbb{R}^n \mid Ax \leq b, x \geq 0\};$$

see Proposition 11.6. Hence, we can first find the solution x^* of (5.2) and afterwards find p^* as a solution of the linear programming problem:

$$\begin{array}{c} \min \to b^T p \\ \text{subject to} \\ A^T p \geq c, p \geq 0 \end{array}$$

where $c = c(x^*)$ is fixed.

Exercise 5.1. By using the results of Section 11.2, find existence and uniqueness conditions for the Cassel-Wald model.

5.2 General price equilibrium models

A great number of economic models are adjusted to investigating the conditions which balance supply and demand of commodities, i.e., they are essentially equilibrium models. As a rule, the concept of economic equilibrium can be written in terms of a complementarity relation between the price and the excess demand for each commodity. Therefore, these economic equilibrium models can be written as complementarity (or variational inequality) problems. To illustrate this assertion, we now describe one of the most general economic models originated by *L. Walras*; see *Walras* (1874).

So, it is assumed that our economy deals in n commodities and that there are m economics agents dealing with these commodities. Let $M = \{1, \ldots, m\}$. We divide M into two subsets M_s and M_c which correspond to sectors (producers) and consumers, respectively. Given a price vector $p \in \mathbb{R}^n_+$, the j-th producer determines his supply $S_j(p) \in \mathbb{R}^n_+$ and the i-th consumer determines his demand $D_i(p) \in \mathbb{R}^n_+$. For simplicity, we suppose the mappings S_j and D_i are single-valued. Set

$$S(p) = \sum_{j \in M_s} S_j(p), D(p) = \sum_{i \in M_c} D_i(p).$$

Then we can define the excess demand mapping

$$E(p) = D(p) - S(p).$$

A vector p^* is said to be an *equilibrium price* if it satisfies the following conditions:

$$p^* \in \mathbb{R}^n_+, \quad -E(p^*) \in \mathbb{R}^n_+, \quad (p^*)^T E(p^*) = 0; \tag{5.3}$$

i.e. the equilibrium problem obviously coincides with complementarity problem (1.16), where $G = -E$, $X = \mathbb{R}^n_+$. Therefore, one could derive the existence and uniqueness conditions for this problem directly from those for general complementarity problems and variational inequalities. Moreover, a number of additional existence and uniqueness results, essentially exploiting features of economic equilibrium models, have been obtained.

We now present a specialization of this very general class of economic equilibrium problems, which can be viewed as some extension of the model suggested by *H. Scarf*; see *Scarf and Hansen* (1973). The model also considers an n-commodity market under perfect competition and includes production with linear technology and consumption. Given prices $p \in \mathbb{R}^n_+$, the total demand of the consumers is also defined as $D(p)$, whereas the supply of the j-th producer is defined as $S_j(p) = x_j a^j$, where a^j is the j-th row of the $l \times n$ technology matrix A and x_j is the activity level of the j-th producer, which can be determined as a solution to the one-dimensional optimization problem:

$$\max_{\alpha \geq 0} \to \{\alpha p^T a^j - f_j(\alpha)\}, \tag{5.4}$$

where f_j is the cost function of the j-th producer, which is supposed to be differentiable and dependent of the activity level of the j-th producer, for $j = 1, \ldots, l$. We can replace (5.4) with the optimality conditions

$$x_j \geq 0, [c_j(x_j) - p^T a^j] \geq 0, x_j[c_j(x_j) - p^T a^j] = 0;$$

where $c_j(x_j) = f_j'(x_j)$.

Exercise 5.2. Derive the above optimality conditions for problem (5.4).

Then the total supply of producers is determined by the conditions

$$S(p) = A^T x, \quad x \geq 0, c(x) - Ap \geq 0, x^T[c(x) - Ap] = 0.$$

Suppose that the initial endowments of consumers are represented by a non-negative vector r. Then the excess demand is given by $E(p) = D(p) - S(p) - r$. Taking into account (5.3), we conclude that the equilibrium of this model will be represented by a pair (p^*, x^*) such that

$$p^* \geq 0, r + A^T x^* - D(p^*) \geq 0, (p^*)^T[r + A^T x^* - D(p^*)] = 0; \quad (5.5)$$

and

$$x^* \geq 0, c(x^*) - Ap^* \geq 0, (x^*)^T[c(x^*) - Ap^*] = 0. \quad (5.6)$$

Observe that the known Scarf model corresponds to the case where $c \equiv 0$ (i.e. f is a constant function). Evidently, this model can be viewed as a particular case of problem (4.11).

Following this approach, we can derive the equilibrium model (4.11) as a particular case of this very general class of Walrasian equilibrium models, i.e. we present the reverse derivation process. Let us consider an m-factor market under perfect competition, which includes supply and industrial demand. Given factor prices $y \in \mathbb{R}^m_+$, the total supply is defined as $b(y)$, whereas the total demand of the producers is defined as $d(y)$. A vector $y^* \in \mathbb{R}^m$ is said to be an *equilibrium factor price* if it satisfies the following conditions:

$$y^* \geq 0; \ b(y^*) - d(y^*) \geq 0, \ (y^*)^T[b(y^*) - d(y^*)] = 0. \quad (5.7)$$

Suppose that the factors are used for producing n commodities for a fixed time period, c_j denoting the price of the j-th commodity and a_{ij} denoting the amount of the i-th factor for producing one unit of the j-th commodity. Next, x_j denotes the unknown output of the j-th commodity and $f_j(x_j)$ denotes the value of the income function of this commodity, i.e. $c_j(x_j) = f'_j(x_j)$ can be viewed as the marginal income, which is dependent only of the output of the j-th commodity. The output x_j of the j-th commodity yields the factor demand $x_j A^j$, where A^j is the j-th column of the $m \times n$ technology matrix A, and the optimal output can be determined as a solution to the one-dimensional optimization problem:

$$\min_{\alpha \geq 0} \to \{\alpha y^T A^j - f_j(\alpha)\},$$

i.e. we have

$$x_j \geq 0, y^T A^j - c_j \geq 0, x_j(y^T A^j - c_j) = 0.$$

Since $c = c(x)$, the total factor demand is determined by the conditions

$$d(y) = Ax, \quad x \geq 0, A^T y - c(x) \geq 0, x^T [A^T y - c(x)] = 0.$$

Combining these conditions with (5.7), we conclude that the equilibrium of this model will be represented by a pair (x^*, y^*) such that

$$\begin{array}{ll} x^* \geq 0, & y^* \geq 0; \\ A^T y^* - c(x^*) \geq 0, & b(y^*) - Ax^* \geq 0; \\ (x^*)^T [A^T y^* - c(x^*)] = 0, & (y^*)^T [b(y^*) - Ax^*] = 0. \end{array}$$

Clearly, these conditions coincide with (4.11). Similarly, the fixed vectors b and c yield the optimality conditions for linear programming problems.

5.3 Spatial price equilibrium models

In these models, which are generalizations of transportation optimization problems, the spatial location of economic agents (markets) is taken into account.

In the *single commodity market model*, there are l agents (sites) connected by transport communications. Let p_i denote the price of the commodity at site i, and f_{ij} be the export from site i to site j. If the prices at all sites $p = (p_1, \ldots, p_l)^T$ and the export vector $f = (f_{11}, \ldots, f_{ll})^T$ are given, we can determine the cost of shipping one unit of the commodity from site i to site j, which is denoted by c_{ij}, and the capacity (or excess demand) d_i of site i (if $d_i > 0$, it is a consuming market; if $d_i < 0$, then it is a supplying market). In general, c and d are some mappings, i.e. $c = c(f)$ and $d = d(p)$.

The equilibrium conditions in this model have the form of a system of complementarity problems:

$$f_{ij}^* \geq 0, \quad p_i^* \geq 0,$$

$$c_{ij}(f^*) - p_j^* + p_i^* \geq 0, \sum_{j=1}^{l} f_{ji}^* - \sum_{j=1}^{l} f_{ij}^* - d_i(p^*) \geq 0, \tag{5.8}$$

$$f_{ij}^* \left(c_{ij}(f^*) - p_j^* + p_i^* \right) = 0, p_i^* \left(\sum_{j=1}^{l} f_{ji}^* - \sum_{j=1}^{l} f_{ij}^* - d_i(p^*) \right) = 0,$$

for all $i, j = 1, \ldots, l$. The first series of constraints in (5.8) is obvious in view of the meaning of f_{ij}^* and p_i^*. The second series ensures the absence of profit in shipping the commodity from site i to site j and imposes restrictions on import-export volumes. The third series of constraints shows that shipping a positive volume of the commodity from site i to site j is possible only if the incomes are equal and that a positive commodity price at site i is possible only if there is a balance between the import-export volume and the capacity at the given site. Similarly, if the vectors c and d are fixed, then problem (5.8) can be replaced with the following pair of transportation type problems:

$$
\begin{array}{ll}
\min \to \sum_{i=1}^{l}\sum_{j=1}^{l} c_{ij} f_{ij} & \max \to \sum_{i=1}^{l} d_i p_i \\
\sum_{j=1}^{l} f_{ji} - \sum_{j=1}^{l} f_{ij} \ge d_i & c_{ij} - p_j + p_i \ge 0 \\
\text{for} \quad i = 1,\dots,l; & \text{for} \quad i,j = 1,\dots,l; \\
f_{ij} \ge 0 \ \text{for} \ i,j = 1,\dots,l; & p_i \ge 0 \ \text{for} \ i = 1,\dots,l.
\end{array}
$$

It is clear that these problems are mutually dual in the sense of linear programming theory.

Exercise 5.3. Show that the transportation problems are particular cases of (4.1′), (4.2′) and (4.3′), (4.4′).

Since problem (5.8) is a system of complementarity problems, it can be rewritten in the form of a system of variational inequalities: Find $f^* \ge 0$ and $p^* \ge 0$ such that

$$\sum_{i=1}^{l}\sum_{j=1}^{l}(c_{ij}(f^*) - p_j^* + p_i^*)(f_{ij} - f_{ij}^*) \ge 0 \quad \forall f \ge 0,$$

$$\sum_{i=1}^{l}\left[\sum_{j=1}^{l} f_{ji}^* - \sum_{j=1}^{l} f_{ij}^* - d_i(p^*)\right](p_i - p_i^*) \ge 0 \quad \forall p \ge 0.$$

Exercise 5.4. By using the results of Section 11.2, find existence and uniqueness conditions for the model with fixed capacities.

We now give a *network formulation of the spatial price equilibrium model.*

The model is determined on a transportation network with the set of nodes N and the set of arcs A. For each node i, y_i denotes the price of a homogeneous commodity and $d_i(y)$ denotes the excess demand at this node where $y = (y_i)_{i\in N}$. For each arc $a \in A$, f_a denotes the flow and $c_a(f)$ denotes the transportation cost for shipping the commodity for this arc, where $f = (f_a)_{a\in A}$. Next, we denote by W the set of all origin-destination pairs in the net, then P_w denotes the set of paths joining pair w and $P = \bigcup_{w\in W} P_w$ denotes the set of all the paths. For each path $p \in P$, x_p denotes the flow and $C_p(x)$ denotes the transportation cost for this path, where $x = (x_p)_{p\in P}$. Set $C(x) = (C_p(x))_{p\in P}$ and $c(f) = (c_a(f))_{a\in A}$, then clearly

$$f = Bx \quad \text{and} \quad C(x) = B^T c(f), \tag{5.9}$$

where B is the arc-path incidence matrix, i.e. $B = (b_{ap})$,

$$b_{ap} = \begin{cases} 1 & \text{if path } p \text{ involves arc a,} \\ 0 & \text{otherwise.} \end{cases}$$

A flow-price pattern (f^*, y^*) is said to be an equilibrium if it satisfies the following conditions:

$$y_i^* \geq 0,$$

$$\sum_{w=(k,i)\in W} \sum_{p\in P_w} x_p^* - \sum_{w=(i,j)\in W} \sum_{p\in P_w} x_p^* - d_i(y^*) \geq 0, \tag{5.10}$$

$$y_i^* \left[\sum_{w=(k,i)\in W} \sum_{p\in P_w} x_p^* - \sum_{w=(i,j)\in W} \sum_{p\in P_w} x_p^* - d_i(y^*) \right] = 0 \ \forall i \in N;$$

and

$$\begin{aligned} & x_p^* \geq 0, \\ & y_i^* - y_j^* + C_p(x^*) \geq 0, \\ & x_p^* \left[y_i^* - y_j^* + C_p(x^*) \right] = 0 \quad \forall p \in P_w, \quad \forall w = (i,j) \in W; \end{aligned} \tag{5.11}$$

where $f^* = Bx^*$. Conditions (5.10) represent equilibrium between input-output flows and prices at each market, whereas conditions (5.11) represent equilibrium between export flows and profits of shipping for each pair of origin-destination markets. Since these conditions are obviously complementarity problems, they can be equivalently rewritten as the system of variational inequalities: Find $x^* \geq 0$ and $y^* \geq 0$ such that

$$\sum_{i\in N} \left[\sum_{w=(k,i)\in W} \sum_{p\in P_w} x_p^* - \sum_{w=(i,j)\in W} \sum_{p\in P_w} x_p^* - d_i(y^*) \right] (y_i - y_i^*) \geq 0$$
$$\forall y_i \geq 0, i \in N;$$

and

$$\sum_{w\in W} \sum_{p\in P_w} \left[(y_i^* - y_j^*) + C_p(x^*) \right] (x_p - x_p^*) \geq 0$$
$$\forall x_p \geq 0 \quad p \in P_w, w \in W.$$

These models can be extended to the multicommodity case.

Let us consider a generalization of the *spatial price equilibrium model for the dynamic case*, i.e., when the equilibrium trajectory of the entire system on a given time interval is analyzed.

The main difference, in addition to the inclusion of a time parameter, is in the allowance for the cost of the commodity storage. For simplicity, we consider a discrete time; i.e., we divide the time interval into subintervals $t = 1, 2, \ldots, T$. Let $f_{ij,t}$ be the supply from site i to site j at time interval t, $v_{i,t}$ be the volume of the commodity stored at the i-th site at the time intervals from $(t-1)$-th to t-th, and $p_{i,t}$ be the price of the commodity at the i-th site during time interval t. As before, there are l agents (sites or markets), and the supply from the i-th to the j-th site at any interval of time is non-negative for all $i, j = 1, \ldots, l$. If the vectors $f = (f_{11,1}, \ldots, f_{ll,T})^T$, $v = (v_{1,1}, \ldots, v_{l,T})^T$ and $p = (p_{1,1}, \ldots, p_{l,T})^T$ are known, one can determine the transportation cost $c_{ij,t}$ of a unit commodity from the i-th to the j-th site at the t-th time interval, the cost $r_{i,t}$ of storing a unit commodity at the i-th site at the time intervals from $(t-1)$-th to t-th, and the capacity $d_{i,t}$ of the i-th site during time interval t. We suppose that c, r, and d are mappings; i.e., $c_{ij,t} = c_{ij,t}(f)$, $r_{i,t} = r_{i,t}(v)$, $d_{i,t} = d_{i,t}(p)$. This means that the transportation cost of a unit of the commodity between two sites can depend on the volumes of transportation, the cost of storing a unit of the commodity at each site can depend on the amount of stored goods, and that its capacity can depend on the prices at the sites (for each interval of time). A trajectory is called equilibrium if it is defined by the vectors f^*, v^* and p^* such that

$$\begin{aligned} f^*_{ij,t} \geq 0, c_{ij,t}(f^*) + p^*_{i,t} - p^*_{j,t} \geq 0, \\ f^*_{ij,t}[c_{ij,t}(f^*) + p^*_{i,t} - p^*_{j,t}] = 0; \end{aligned} \tag{5.12}$$

for all $i, j = 1, \ldots, l$ and $t = 1, \ldots, T$;

$$\begin{aligned} v^*_{i,t} \geq 0, \ r_{i,t}(v^*) + p^*_{i,t-1} - p^*_{i,t} \geq 0, \\ v^*_{i,t}\left[r_{i,t}(v^*) + p^*_{i,t-1} - p^*_{i,t}\right] = 0 \end{aligned} \tag{5.13}$$

for all $i = 1, \ldots, l$ and $t = 1, \ldots, T$ ($p^*_{i,0} = 0$ are fixed for all $i = 1, \ldots, l$);

$$\begin{aligned} & p^*_{i,t} \geq 0, \\ & \sum_{j=1}^{l} f^*_{ji,t} - \sum_{j=1}^{l} f^*_{ij,t} + v^*_{i,t} - v^*_{i,t+1} - d_{i,t}(p^*) \geq 0, \\ & p^*_{i,t}\left[\sum_{j=1}^{l} f^*_{ji,t} - \sum_{j=1}^{l} f^*_{ij,t} + v^*_{i,t} - v^*_{i,t+1} - d_{i,t}(p^*)\right] = 0; \end{aligned} \tag{5.14}$$

for all $i = 1, \dots, l$ and $t = 1, \dots, T$ ($v^*_{i,T+1} = 0$ are fixed for all $i = 1, \dots, l$).

It is easy to see that relations (5.12) represent the usual complementarity conditions between export flows and profits of shipping for each pair of sites. Next, conditions (5.13) indicate that it is not advantageous to store commodities and a positive amount can be stored only if the prices at the t-th and $(t+1)$-th intervals with account for the cost of storage are balanced. Similarly, (5.14) imposes constraints on the import-export and shows that a positive commodity price at the i-th site at time t is possible only if there is an exact balance between the import-export volume (with account for the storage cost) and the capacity at this site.

As in the static model, system (5.12)–(5.14) is reduced to a system of variational inequalities. More precisely, problem (5.12) is equivalent to the following one: Find a point $f^* \geq 0$ such that

$$\sum_{t=1}^{T} \sum_{i=1}^{l} \sum_{j=1}^{l} \left[c_{ij,t}(f^*) + p^*_{i,t} - p^*_{j,t}\right] (f_{ij,t} - f^*_{ij,t}) \geq 0 \qquad \forall f_{ij,t} \geq 0, \tag{5.15}$$

for all $i, j = 1, \dots, l$ and $t = 1, \dots, T$. The complementarity problem (5.13) is reduced to the following variational inequality: Find a point $v^* \geq 0$ such that

$$\sum_{t=1}^{T} \sum_{i=1}^{l} \left[r_{i,t}(v^*) + p^*_{i,t-1} - p^*_{i,t}\right] (v_{i,t} - v^*_{i,t}) \geq 0 \qquad \forall v_{i,t} \geq 0 \tag{5.16}$$

for all $i = 1, \dots, l$ and $t = 1, \dots, T$. Due to the independence of the variables f and v, we can replace system (5.15), (5.16) by the single variational inequality: Find points $w^* \geq 0$ and $v^* \geq 0$ such that

$$\begin{aligned} &\sum_{t=1}^{T} \sum_{i=1}^{l} \sum_{j=1}^{l} \left[c_{ij,t}(f^*) + p^*_{i,t} - p^*_{j,t}\right] (f_{ij,t} - f^*_{ij,t}) \\ &+ \sum_{t=1}^{T} \sum_{i=1}^{l} \left[r_{i,t}(v^*) + p^*_{i,t-1} - p^*_{i,t}\right] (v_{i,t} - v^*_{i,t}) \geq 0 \\ &\forall f_{ij,t} \geq 0 \quad \text{and} \quad \forall v_{i,t} \geq 0 \end{aligned} \tag{5.17}$$

for all $i, j = 1, \dots, l$ and $t = 1, \dots, T$. Furthermore, the complementarity problem (5.14) is equivalent to the following variational inequality: Find a point $p^* \geq 0$ such that

$$\sum_{t=1}^{T}\sum_{i=1}^{l}\left[\sum_{j=1}^{l} f^*_{ji,t} - \sum_{j=1}^{l} f^*_{ij,t} + v^*_{i,t} - v^*_{i,t+1}\right.$$
$$\left. - d_{i,t}(p^*)\right](p_{i,t} - p^*_{i,t}) \geq 0 \quad \forall p_{i,t} \geq 0 \tag{5.18}$$

for all $i = 1, \ldots, l$ and $t = 1, \ldots, T$.

There exist various modifications and extensions of spatial price equilibrium models; see *Harker* (1985), *Miller, Friesz, and Tobin* (1996), and *Nagurney* (1999).

5.4 Imperfectly competitive equilibrium models

We now consider the problem of finding market equilibria for the case of a few economic agents (producers). It means that actions of each separate agent can change the state of the whole system. These oligopolistic equilibrium models originated by *A. Cournot* belong to imperfectly competitive systems; see *Cournot* (1838).

In the classical oligopoly model, it is assumed that there are n firms supplying a homogeneous product and that the price p depends on its quantity σ, i.e. $p = p(\sigma)$ is the inverse demand function. In other words, $p(\sigma)$ is the price at which consumers will purchase a quantity σ. Next, the value $h_i(x_i)$ represents the i-th firm total cost of supplying x_i units of the product. If each i-th firm supplies x_i units of the product, then the total supply in the market is defined by

$$\sigma_x = \sum_{i=1}^{n} x_i,$$

and the i-th firm's profit is defined by

$$f_i(x) = x_i p(\sigma_x) - h_i(x_i),$$

where $x = (x_1, x_2, \ldots, x_n)^T$. Of course, each output level is nonnegative, i.e., $x_i \geq 0$ for $i = 1, \ldots, n$. Naturally, each firm seeks to maximize its own profit by choosing the corresponding production level. However, since the profit of each firm is dependent of outputs of all the firms, whose interests may be rather different, we can consider this problem as a non-cooperative game of n players, where the i-th player has the strategy set $\mathbb{R}_+$ and the utility function $f_i(x)$. Therefore, in order to define a solution in this market structure we use the *Nash* equilibrium concept for non-cooperative games; see e.g. *Okuguchi and Szidarovszky* (1990) and *Aubin* (1998). By definition, a nonnegative vector of output levels $x^* = (x_1^*, x_2^*, \dots, x_n^*)^T$ is said to constitute a Nash equilibrium solution for the oligopolistic market, provided x_i^* maximizes the profit function f_i of the i-th firm given that the other firms produce quantities x_j^*, $j \neq i$, for each $i = 1, \dots, n$.

That is, for $x^* = (x_1^*, x_2^*, \dots, x_n^*)^T$ to be a Nash equilibrium, x_i^* must be an optimal solution to the problem

$$\max_{x_i \geq 0} \to \{x_i p(x_i + \sigma_i^*) - h_i(x_i)\}, \tag{5.19}$$

where $\sigma_i^* = \sum\limits_{j=1, j \neq i}^{n} x_j^*$ for each $i = 1, \dots, n$. At the same time, this problem can be transformed into an equivalent variational inequality or complementarity problem if each i-th profit function f_i is concave in x_i. This assumption conforms to the usually accepted economic behavior and implies that (5.19) is a concave maximization problem. In addition, we assume for simplicity that the price function $p(\sigma)$ and all the cost functions $h_i(x_i)$ are continuously differentiable. Namely, set $G_i(x) = -p(\sigma_x) - x_i p'(\sigma_x) + h_i'(x_i)$ for $i = 1, \dots, n$, thus defining the mapping $G : \mathbb{R}_+^n \to \mathbb{R}^n$. Under the assumptions above, each optimization problem (5.19) is equivalent to the complementarity problem:

$$x_i^* \geq 0, G_i(x^*) \geq 0, x_i^* G_i(x^*) = 0; \tag{5.20}$$

or to the variational inequality: Find $x_i^* \geq 0$ such that

$$G_i(x^*)(x_i - x_i^*) \geq 0 \quad \forall x_i \geq 0;$$

for each $i = 1, \dots, n$. However, this system of partial variational inequality problems is equivalent to the usual variational inequality: Find $x^* \geq 0$ such that

$$(x - x^*)^T G(x^*) \geq 0 \quad \forall x \geq 0; \tag{5.21}$$

see also Section 7.1.

It should be noticed that many problems in economics and social sciences may be formulated as game equilibrium models; see e.g. *von Neumann and Morgenstern* (1953), *Moulin* (1981), *Okuguchi and Szidarovszky* (1990), and *Aubin* (1998). Therefore, following the above approach, they can be investigated with the help of the corresponding equivalent variational inequality problem.

We illustrate this model with some examples. We start from the case considered by A. Cournot.

Example 5.1. Let $n = 2$, the functions p and h_i be affine, i.e.,

$$\begin{array}{rcl} p(\sigma) & = & \alpha - \beta\sigma, \alpha \geq 0, \beta > 0; \\ h_i(x_i) & = & \gamma x_i + \delta, \gamma \geq 0, \delta \geq 0 \quad \text{for} \quad i = 1, 2. \end{array}$$

It means that both the producers have equal opportunities, i.e. we have obtained the classical Cournot duopoly. Suppose that $\alpha > \gamma$ (i.e. the starting price is greater than the marginal expenses) and set $\tau = (\alpha - \gamma)/\beta$. Then we can drop the nonnegativity constraints in (5.20) and consider the system of equations

$$G_i(x) = 0 \quad \text{for} \quad i = 1, 2;$$

or, equivalently,

$$\begin{cases} 2x_1 + x_2 & = \tau \\ x_1 + 2x_2 & = \tau. \end{cases}$$

This system has clearly the unique and symmetric solution:

$$x_1^* = x_2^* = \tau/3.$$

The solution can be explained from the game theoretic point of view (see Figure 5.1). In fact, if the output x_2 of the second player is known, the best choice of the first player can be found from the one-dimensional equation: $p(x_1 + x_2) + x_1 p'(x_1 + x_2) - h_1'(x_1) = 0$ or, equivalently, $2x_1 + x_2 = \tau$. It follows that the optimal reaction of the first player is $x_1(x_2) = (\tau - x_2)/2$. Similarly, if the output x_1 of the first player is known, the best choice of the second player can be found from the equation: $x_1 + 2x_2 = \tau$, i.e., the optimal reaction of the second player is $x_2(x_1) = (\tau - x_1)/2$. The intersection of these reaction lines corresponds to the equilibrium point x^*, where the reactions coincide with outputs.

Now we consider a somewhat more general case.

Exercise 5.5. Let the functions p and h_i be affine, i.e.,

$$\begin{array}{rcl} p(\sigma) & = & \alpha - \beta\sigma, \alpha \geq 0, \beta > 0; \\ h_i(x_i) & = & \gamma_i x_i + \delta_i, \gamma_i \geq 0, \delta_i \geq 0 \quad \text{for} \quad i = 1, \dots, n. \end{array}$$

Then,

$$f_i(x) = x_i(\alpha - \beta\sigma_x) - \gamma_i x_i - \delta_i.$$

Find the value of $G(x)$ and show that the oligopolistic equilibrium model becomes equivalent to the problem of minimizing a strongly convex quadratic function subject to the nonnegativity constraints.

Figure 5.1:

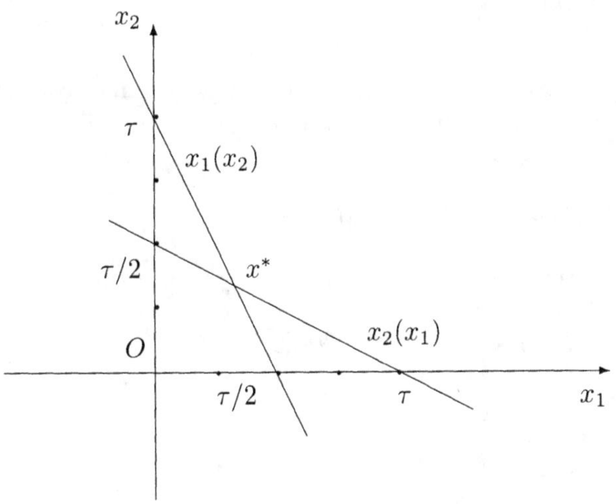

Another approach to modeling imperfect competition markets was proposed by *J. Bertrand.* Following this approach, one also considers a market where n firms supply a homogeneous product, but they announce prices rather than volumes. More precisely, if each i-th firm indicates its price p_i, $i = 1, \dots, n$, then consumers are able to determine the demand quantities q_i for each $i = 1, \dots, n$ and $q_i = q_i(p)$ where $p = (p_1, \dots, p_n)^T$. Then $h_i(q_i)$ represents the i-th firm's total cost of supplying q_i units of the product, and the i-th firm's profit is defined by

$$f_i(p) = p_i q_i(p) - h_i[q_i(p)] \quad \text{for} \quad i = 1, \dots, n.$$

Again, the profit of each firm depends on prices of all the firms in general and we can use the Nash equilibrium concept in the non-cooperative game of n players, where the i-th player has the utility function $f_i(p)$ and the strategy set $\mathbb{R}_+$, as a solution of the *Bertrand oligopoly* model. Usually, each function $q_i(p)$ is supposed to be decreasing in p_i and increasing in other variables, which corresponds to the usually accepted economic behavior of consumers, and, under certain additional assumptions, the problem can be also reduced to a complementarity problem or a variational inequality.

In fact, $p^* = (p_1^*, \dots, p_n^*)^T$ is a Nash equilibrium point if and only if p_i^*

is a solution to the optimization problem

$$\begin{array}{ll}\max\limits_{p_i\geq 0}\rightarrow & \{p_iq_i(p_1^*,\ldots,p_{i-1}^*,p_i,p_{i+1}^*,\ldots,p_n^*)\\ & -h_i\left[q_i(p_1^*,\ldots,p_{i-1}^*,p_i,p_{i+1}^*,\ldots,p_n^*)\right]\}\\ & \text{for}\quad i=1,\ldots,n.\end{array} \tag{5.22}$$

If f_i is differentiable and convex in p_i, (5.22) becomes a system of concave differentiable maximization problems and equivalent to the complementarity problem

$$p_i^*\geq 0, G_i(p^*)\geq 0, p_i^*G_i(p^*)=0 \quad \text{for}\quad i=1,\ldots,n$$

(cf. (5.20)) or to the variational inequality: Find $p^*\geq 0$ such that

$$(p-p^*)^TG(p^*)\geq 0 \quad \forall p\geq 0$$

(cf. (5.21)), where

$$G_i(p)=-\frac{\partial f_i(p)}{\partial p_i} \quad \text{for}\quad i=1,\ldots,n;$$

on account of Corollary 11.2 and Proposition 7.1.

Example 5.2. Let $q_i(p)=\alpha_i\left(\sum\limits_{j\neq i}p_j/p_i\right)-\beta_i, h_i(t)=\gamma_it+\delta_i$ with $\alpha_i,\beta_i,\gamma_i,\delta_i>0$ for $i=1,\ldots,n$. Then

$$f_i(p)=\alpha_i\sum_{j\neq i}p_j-\beta_ip_i-\gamma_i\left[\alpha_i\left(\sum_{j\neq i}p_j/p_i\right)-\beta_i\right]-\delta_i$$

is concave and differentiable in p_i on $\mathrm{int}\mathbb{R}^n_+$. Hence we can replace the Nash equilibrium problem by the complementarity problem (or variational inequality) above.

In the case when the system

$$G_i(p)=0 \quad \text{for}\quad i=1,\ldots,n \tag{5.23}$$

has positive solutions, they are clearly equilibrium points. This approach may be useful if G is not defined at points including zero coordinates.

Exercise 5.6. Set $n=2, q_i(p)=\alpha p_{3-i}/p_i-\beta, h_i(t)=\gamma t+\delta$ with $\alpha,\beta,\gamma,\delta>0$ for $i=1,2$ and find the equilibrium point via solution of system (5.23).

Figure 5.2: $p^* = p_3$

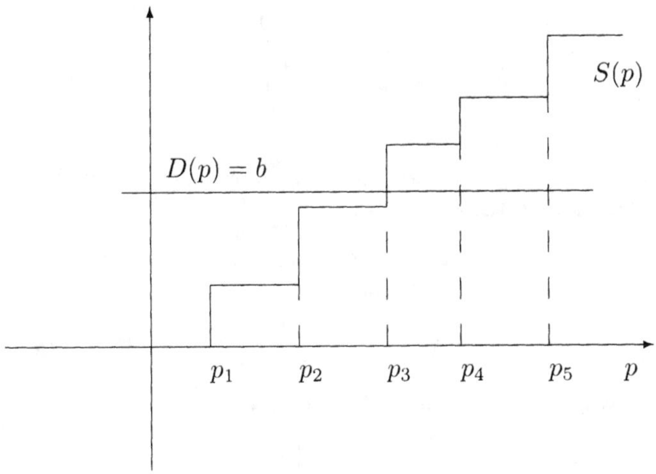

There exist more general oligopolistic equilibrium models which take into account the spatial locations of producers and markets and involve many commodities; see e.g. *Miller, Friesz, and Tobin* (1996), *Nagurney* (1999), and *Facchinei and Pang* (2003).

We now describe another class of imperfectly competitive equilibrium models related to *auction type markets.* More precisely, the first model represents an auction of n traders where the total bid is fixed and denoted by b. In the simplified formulation, each i-th trader indicates his maximal offer β_i and price p_i, thus determining the total supply function. Then the equilibrium price p^* can be found very easily. Without loss of generality we suppose that $i < j$ implies $p_i \leq p_j$. It suffices to find k such that

$$\sum_{i<k} \beta_i < b \text{ and } \sum_{i\leq k} \beta_i \geq b,$$

and afterwards set $p^* = p_k$, $x_i = \beta_i$ if $i < k$ and $x_k = \min\{\beta_k, b - \sum\limits_{i<k} \beta_i\}$; see Figure 5.2.

However, we are interested in more complicated models which involve prices depending on offer values. If x_i is the offer value of trader i and $p_i : \mathbb{R}^n_+ \to \mathbb{R}_+$ is his inverse supply function whose value $p_i(x)$ may depend on all offers, then the equilibrium conditions are written as follows:

$$x^* \in X, p_i(x^*) \left\{ \begin{array}{ll} = \lambda & \text{if } x_i^* > 0, \\ \geq \lambda & \text{if } x_i^* = 0, \end{array} \right. \quad \text{for } i = 1, \ldots, n, \tag{5.24}$$

where

$$X = \left\{ x \in \mathbb{R}_+^n \;\middle|\; \sum_{i=1}^{n} x_i = b \right\},$$

λ is the (unknown) auction clearing price.

However, conditions (5.24) are equivalent to a variational inequality, as the following proposition states.

Proposition 5.1. *A point $x^* \in \mathbb{R}^n$ satisfies conditions (5.24) if and only if it solves the problem: Find $x^* \in X$ such that*

$$(x - x^*)^T p(x^*) \geq 0 \quad \forall x \in X. \tag{5.25}$$

Proof. Let x^* satisfy (5.24). Take any $x \in X$ and set

$$I_0 = \{i \mid x_i^* = 0\} \quad \text{and} \quad I_+ = \{i \mid x_i^* > 0\}.$$

Then we have

$$\begin{aligned} (x - x^*)^T p(x^*) &= \sum_{i=1}^{n} (x_i - x_i^*) p_i(x^*) \\ &= \sum_{i \in I_0} p_i(x^*)(x_i - 0) + \sum_{i \in I_+} \lambda (x_i - x_i^*) \\ &\geq \lambda \sum_{i=1}^{n} (x_i - x_i^*) = \lambda (b - b) = 0, \end{aligned}$$

i.e. x^* solves (5.25). Conversely, let x^* solve (5.25), then $x^* \in X$. Take any indices k and l such that $p_k(x^*) > p_l(x^*)$ and determine the point $\tilde{x}$ by the rule:

$$\tilde{x}_i = \left\{ \begin{array}{ll} x_i^* & \text{if } i \neq k \text{ or } i \neq l, \\ x_l^* + x_k^* & \text{if } i = l, \\ 0 & \text{if } i = k, \end{array} \right. \quad \text{for } i = 1, \ldots, n.$$

Then $\tilde{x} \in X$ and

$$\begin{aligned} 0 &\leq (\tilde{x} - x^*)^T p(x^*) = p_k(x^*)(\tilde{x}_k - x_k^*) + p_l(x^*)(\tilde{x}_l - x_l^*) \\ &= x_k^* \left[p_l(x^*) - p_k(x^*) \right] \leq 0, \end{aligned}$$

hence $x_k^* = 0$. It means that conditions (5.24) are fulfilled. □

Observe that formulation (5.25) allows us to find the unknown auction price p^* by setting

$$p^* = \min_{i=1,\dots,n} p_i(x^*).$$

This model appears close to the costless migration equilibrium models; see *Nagurney* (1999). Here its statement seems very natural.

Exercise 5.7. By using the theorems of Section 11.2, find existence and uniqueness results for problem (5.25).

This model can be extended in several directions. For instance, let us consider the case where each i-th trader has the minimal and maximal offer volumes, denoted by α_i and β_i, respectively. Then conditions (5.24) are replaced by the following:

$$x^* \in \tilde{X}, p_i(x^*) \left\{ \begin{array}{ll} \geq \lambda & \text{if } x_i^* = \alpha_i, \\ = \lambda & \text{if } x_i^* \in (\alpha_i, \beta_i), \\ \leq \lambda & \text{if } x_i^* = \beta_i, \end{array} \right. \qquad (5.26)$$
$$\text{for } i = 1, \dots, n,$$

where

$$\tilde{X} = \left\{ x \in \mathbb{R}^n \;\middle|\; \sum_{i=1}^{n} x_i = b, \ \alpha_i \leq x_i \leq \beta_i \ \text{ for } \ i = 1, \dots, n \right\}.$$

Nevertheless, these conditions are equivalent to the following variational inequality: Find $x^* \in \tilde{X}$ such that

$$(x - x^*)^T p(x^*) \geq 0 \quad \forall x \in \tilde{X} \qquad (5.27)$$

(cf. (5.25)). This property can also be deduced from the fact that relations (5.26) represent optimality conditions for constrained problems; see Chapter 11.

Exercise 5.8. Deduce the equivalence of (5.26) and (5.27) from Proposition 11.7. By using the theorems of Section 11.2, find existence and uniqueness results for problem (5.27).

Due to equivalent formulations (5.25) and (5.27), the equilibrium quantities and clearing price can be found rather easily.

Exercise 5.9. Find solutions of problem (5.26) when $n = 2$, $\alpha_1 = 1$, $\beta_1 = 2$, $\alpha_2 = 1$, $\beta_2 = 3$, $p_1(x_1) = 2x_1 - 1$, $p_2(x_2) = 0.5x_2 + 1$ for $b = 3$ and for $b = 5$ and give their graphical illustration.

Observe that auctions of n buyers with the fixed total offer can be considered similarly. It is sufficient to utilize the opposite inequality signs in all the relations (5.24)–(5.27).

We can extend this approach for modeling the auction market involving both traders and buyers. Let x_i denote the offer value of trader i for $i = 1, \dots, n$ and let y_j denote the bid value of buyer j for $j = 1, \dots, m$. Given vectors $x \in \mathbb{R}^n$ and $y \in \mathbb{R}^m$, each i-th trader determines his price (inverse supply) $g_i(x, y)$ and each j-th buyer determines his price (inverse demand) $h_j(x, y)$. We define the feasible set

$$Z = \left\{ (x, y) \in \mathbb{R}^n_+ \times \mathbb{R}^m_+ \;\middle|\; \sum_{i=1}^{n} x_i = \sum_{j=1}^{m} y_j \right\} \tag{5.28}$$

and the variational inequality: Find $(x^*, y^*) \in Z$ such that

$$\begin{array}{c} \sum\limits_{i=1}^{n} g_i(x^*, y^*)(x_i - x_i^*) - \sum\limits_{j=1}^{m} h_j(x^*, y^*)(y_j - y_j^*) \geq 0 \\ \forall (x, y) \in Z. \end{array} \tag{5.29}$$

Writing optimality conditions for this problem (see Proposition 11.7) gives $(x^*, y^*) \in Z$,

$$g_i(x^*, y^*) \left\{ \begin{array}{ll} = \lambda & \text{if } x_i^* > 0, \\ \geq \lambda & \text{if } x_i^* = 0, \end{array} \right. \qquad \text{for } i = 1, \dots, n,$$

and

$$h_j(x^*, y^*) \left\{ \begin{array}{ll} = \lambda & \text{if } y_j^* > 0, \\ \leq \lambda & \text{if } y_j^* = 0, \end{array} \right. \qquad \text{for } j = 1, \dots, m,$$

for some λ. Clearly, these conditions define the equilibrium state of this market and λ is precisely the auction clearing price.

Exercise 5.10. By using the theorems of Section 11.2, find the general existence result for problem (5.28), (5.29).

Similarly, we can formulate equilibrium conditions for the case when bids and offers are restricted by upper and lower bounds.

Chapter 6

Transportation and Migration Models

Together with economic applications of equilibrium models we will consider similar models in related fields.

6.1 Network equilibrium models

Flow equilibrium problems in transportation and communication systems constitute rather a new but broad and rapidly developing area of applications of variational inequalities. An essential feature of such problems consists mainly in the fact that they are determined on an oriented graph, each its arc being associated with some flow (for instance, traffic) and some expense (for instance, the time of motion, the time of delay, or cost etc), which depends on the values of arc flows. It is expected that increasing the value of flow for some arc increase the expense for this arc and perhaps for several neighbor arcs, which in turn implies redistribution of flows resulting in some equilibrium state. This means that they are close to spatial price equilibrium models.

There are a great number of various formulations of network equilibrium problems; see e.g. *Patriksson* (1994), *Giannessi and Maugeri* (1995), *Nagurney* (1999), and *Facchinei and Pang* (2003). We first consider a *multicommodity formulation.*

The model is determined on a transportation network given by a set of nodes N and a set of arcs A. We denote by D the subset of destination nodes, $D \subseteq N$. The variable x^l_a denotes the flow on arc a with destination $l \in D$, so that we have

$$x^l = (x^l_a)_{a \in A} \quad \text{and} \quad x = (x^l)_{l \in D}.$$

Similarly, the variable t^l_i denotes the minimal cost to reach destination l from node i, so that we have

$$t^l = (t^l_i)_{i \in N} \quad \text{and} \quad t = (t^l)_{l \in D}.$$

For each pair $(l, i) \in D \times N$, d^l_i denotes the flow demand, i.e. the minimal demand for transportation from node i to node l, which is supposed to be fixed. For each $i \in N$, A^+_i and A^-_i denote the sets of outgoing and incoming arcs at i. Next, for each arc a, c_a denotes the flow cost on this arc, which is dependent of the flow vector

$$f = \sum_{l \in D} x^l.$$

The pair $(\tilde{x}, \tilde{t})$ is said to be in equilibrium if the following conditions hold:

$$\begin{array}{c} \tilde{x}^l_e \geq 0, \quad \tilde{t}^l_i \geq 0; \\ \tilde{t}^l_j - \tilde{t}^l_i + c_e(\tilde{f}) \geq 0, \quad \sum\limits_{a \in A^+_i} \tilde{x}^l_a - \sum\limits_{a \in A^-_i} \tilde{x}^l_a - d^l_i \geq 0; \\ \tilde{x}^l_e \left[\tilde{t}^l_j - \tilde{t}^l_i + c_e(\tilde{f}) \right] = 0, \quad \tilde{t}^l_i \left[\sum\limits_{a \in A^+_i} \tilde{x}^l_a - \sum\limits_{a \in A^-_i} \tilde{x}^l_a - d^l_i \right] = 0; \end{array} \tag{6.1}$$

for all $e = (i, j) \in A, l \in D$ and for all $i \in N, l \in D$, where $\tilde{f} = \sum\limits_{l \in D} \tilde{x}^l$. The first pair of relations represents the non-negativity of flows and costs. The second pair of inequalities in (6.1) means that the difference of minimal costs at any two nodes cannot exceed the flow cost on the corresponding arc and that the flow demand cannot exceed the difference between outgoing and incoming flows. The third pair of relations in (6.1) means that the positive flow on each arc (respectively, the positive minimal cost at each node) yields the equality in the previous series of conditions. Hence, if it is necessary to provide the flow balance at each node:

$$\sum_{a \in A^+_i} \tilde{x}^l_a - \sum_{a \in A^-_i} \tilde{x}^l_a = d^l_i,$$

one has to simply guarantee the positivity of the transportation costs. It is clear that conditions (6.1) determine a nonlinear complementarity problem, hence they may be rewritten in the form of a system of variational inequalities: Find $(\tilde{x}, \tilde{t}) \geq 0$ such that

$$\begin{array}{l} \sum\limits_{a \in A} c_a(\tilde{f})(f_a - \tilde{f}_a) \\ + \sum\limits_{a=(i,j) \in A} \sum\limits_{l \in D} (\tilde{t}^l_j - \tilde{t}^l_i)(x^l_a - \tilde{x}^l_a) \geq 0 \ \forall x \geq 0, \end{array} \tag{6.2}$$

$$\sum_{i\in N}\sum_{l\in D}\left[\sum_{a\in A_i^+}\tilde{x}_a^l - \sum_{a\in A_i^-}\tilde{x}_a^l - d_i^l\right](t_i^l - \tilde{t}_i^l) \geq 0 \quad \forall t \geq 0; \tag{6.3}$$

where $\tilde{f} = \sum_{l\in D} \tilde{x}^l$. Obviously, this system also represents optimality conditions for the variational inequality problem: Find $\tilde{x} \in X$ such that

$$\tilde{f} = \sum_{l\in D} \tilde{x}^l, \quad \sum_{a\in A} c_a(\tilde{f})(f_a - \tilde{f}_a) \geq 0 \quad \forall f \in F,$$

where

$$F = \left\{ f \,\middle|\, f = \sum_{l\in D} x^l, x \in X \right\}$$

and

$$X = \left\{ x \geq 0 \,\middle|\, \sum_{a\in A_i^+} x_a^l - \sum_{a\in A_i^-} x_a^l \geq d_i^l \quad \forall i \in N, \forall l \in D \right\}.$$

We now consider a *path flow formulation* of the same network equilibrium problem.

Let us given a graph with a finite set of nodes N and a set of oriented arcs A which join the nodes so that an arc $a = (i \to j)$ has the origin i and the destination j. Next, among all the pairs of nodes of the graph we extract a subset of pairs W of the form $w = (i \to j)$, where i is the origin node and j is the destination node. Besides, each pair $w \in W$ is associated with a positive number d_w which gives the flow demand from i to j. Denote by P_w the set of paths in the graph which connect the origin and destination, for the pair $w \in W$. Also, denote by x_p the path flow for the path p. Then, the feasible set of flows X can be defined as follows:

$$X = \left\{ x \,\middle|\, \sum_{p\in P_w} x_p = d_w, x_p \geq 0 \; p \in P_w; w \in W \right\}, \tag{6.4}$$

i.e.

$$X = \prod_{w\in W} X_w,$$

where

$$X_w = \left\{ x \,\middle|\, \sum_{p\in P_w} x_p = d_w, x_p \geq 0 \; p \in P_w \right\}.$$

If we set m to be the number of origin-destination pairs and n_w to be the number of paths joining the nodes of the pair w, then the total number of variables in this problem equals

$$n = n_1 \times \ldots \times n_m.$$

Next, if the flow vector x is known, we can determine the value of the arc flow f_l for each arc $l \in A$:

$$f_l = \sum_{w \in W} \sum_{p \in P_w} \alpha_{pl} x_p,$$

where

$$\alpha_{pl} = \begin{cases} 1 & \text{if arc } l \text{ belongs to path } p, \\ 0 & \text{otherwise.} \end{cases}$$

If the values of arc flows are known, one can determine the value of expenses (costs) for each arc as follows:

$$t_l = T_l(f), \tag{6.5}$$

which in general depends on flows for other arcs and uses some mapping T that is defined in the space of flows. Then one can compute the value of expenses for each path p as follows:

$$g_p = G_p(x) = \sum_{l \in A} \alpha_{pl} t_l. \tag{6.6}$$

The feasible flow vector $x^* \in X$ is said to be an equilibrium vector if it satisfies the following conditions:

$$\begin{array}{c} \forall q \in P_w, x_q^* > 0 \Longrightarrow G_q(x^*) = \min_{p \in P_w} G_p(x^*) \\ \text{for all } \; w \in W. \end{array} \tag{6.7}$$

In other words, positive values of flow for any origin - destination pairs must correspond to paths with minimal costs. It is known that the conditions (6.4)–(6.7) can be equivalently rewritten in the form of the variational inequality: Find a point $x^* \in X$ such that

$$(x - x^*)^T G(x^*) \geq 0 \quad \forall x \in X, \tag{6.8}$$

where the inner product is defined in the n-dimensional space of paths joining all the nodes from W.

Theorem 6.1. *A vector x^* solves the problem (6.8) if and only if it satisfies the conditions (6.4)–(6.7).*

Proof. Let x^* be an equilibrium vector. For every $w \in W$, set

$$\lambda_w = \min_{p \in P_w} G_p(x^*),$$
$$P'_w = \{p \in P_w \mid G_p(x^*) = \lambda_w\}, \quad P''_w = P_w \backslash P'_w,$$

and take any feasible flow vector $x \in X$. Then we have

$$(x - x^*)^T G(x^*) = \sum_{w \in W} \sum_{p \in P_w} G_p(x^*)(x_p - x_p^*)$$

$$= \sum_{w \in W} \left[\sum_{p \in P'_w} G_p(x^*)(x_p - x_p^*) + \sum_{p \in P''_w} G_p(x^*) x_p \right]$$

$$\geq \sum_{w \in W} \lambda_w \left[\sum_{p \in P'_w} (x_p - x_p^*) + \sum_{p \in P''_w} x_p \right]$$

$$= \sum_{w \in W} \lambda_w \sum_{p \in P_w} (x_p - x_p^*) = \sum_{w \in W} \lambda_w (d_w - d_w) = 0,$$

i.e. x^* solves variational inequality (6.8).

Conversely, let x^* solve variational inequality (6.8). Choose an arbitrary pair $w \in W$ and arbitrary paths $u, v \in P_w$ such that $G_u(x^*) > G_v(x^*)$. Define the flow vector $\tilde{x}$ as follows:

$$\tilde{x}_p = \begin{cases} x_p^* & \text{if } p \neq u \text{ and } p \neq v, \\ x_u^* + x_v^* & \text{if } p = v, \\ 0 & \text{if } p = u. \end{cases}$$

By construction, $\tilde{x}_p \geq 0$. For each pair $w' \in W$, $w' \neq w$, we have

$$\sum_{p \in P_{w'}} \tilde{x}_p = \sum_{p \in P_{w'}} x_p^* = d_{w'}.$$

Besides,

$$\sum_{p \in P_w} \tilde{x}_p = \sum_{p \neq u, p \neq v} x_p^* + x_u^* + x_v^* = d_w.$$

Therefore, $\tilde{x} \in X$. Next, by definition,

$$\sum_{w \in W} \sum_{p \in P_w} G_p(x^*)(\tilde{x}_p - x_p^*) \geq 0,$$

but

$$\sum_{w \in W} \sum_{p \in P_w} G_p(x^*)(\tilde{x}_p - x_p^*)$$

$$= G_u(x^*)(\tilde{x}_u - x_u^*) + G_v(x^*)(\tilde{x}_v - x_v^*)$$
$$= G_u(x^*)(-x_u^*) + G_v(x^*)x_u^*$$
$$= x_u^*(G_v(x^*) - G_u(x^*)) \leq 0.$$

Therefore, $x_u^* = 0$ and the result follows. □

Exercise 6.1. By using the results of Section 11.2, find existence conditions for this model.

6.2 Migration equilibrium models

We now consider a *migration equilibrium model*, which can be regarded as a somewhat simplified version of the model suggested by *A. Nagurney*; see *Nagurney* (1999).

The model involves a set of nodes (locations) $\mathbf{N}$, for each $i \in \mathbf{N}$, b_i denotes the initial fixed population in location i. Let h_{ij} denote the value of the migration flow from origin i to destination j, and let x_i denote the current population in location i. We can associate with each location i the utility u_i and with each pair of locations i, j the migration cost c_{ij}. Set $\mathbf{x} = (x_i \mid i \in \mathbf{N})$ and $\mathbf{h} = (h_{ij} \mid i, j \in \mathbf{N}, i \neq j)$, then the feasible set can be defined as follows:

$$H = \left\{ (\mathbf{x}, \mathbf{h}) \;\middle|\; \mathbf{h} \geq 0, \quad \sum_{j \neq i} h_{ij} \leq b_i, \right. \\ \left. x_i = b_i + \sum_{j \neq i} h_{ji} - \sum_{j \neq i} h_{ij}, \ \forall i \in \mathbf{N} \right\}. \tag{6.9}$$

Exercise 6.2. Prove that H is bounded.

The rules in (6.9) reflect the conservation of flows and prevent any chain migration. Also, clearly, the migration flow has to be non-negative.

The equilibrium conditions for the scalar migration model are more complicated than those in network equilibrium models. Suppose that the utility depends on the population, i.e. $u_i = u_i(\mathbf{x})$, and that the migration cost depends on the migration flows, i.e. $c_{ij} = c_{ij}(\mathbf{h})$. We say that a pair $(\mathbf{x}^*, \mathbf{h}^*) \in H$ is in equilibrium if

$$u_i(\mathbf{x}^*) - u_j(\mathbf{x}^*) + c_{ij}(\mathbf{h}^*) + \mu_i \left\{ \begin{array}{ll} = 0 & \text{if} \quad h_{ij}^* > 0, \\ \geq 0 & \text{if} \quad h_{ij}^* = 0; \end{array} \right. \tag{6.10}$$

for all $i, j \in \mathbf{N}$ and

$$\mu_i \left\{ \begin{array}{ll} \geq 0 & \text{if} \quad \sum_{s \neq i} h_{is}^* = b_i, \\ = 0 & \text{if} \quad \sum_{s \neq i} h_{is}^* < b_i, \end{array} \right. \tag{6.11}$$

for each $i \in \mathbf{N}$.

The set of equilibrium conditions (6.10), (6.11) can be equivalently rewritten in the form of the variational inequality: Find a pair $(\mathbf{x}^*, \mathbf{h}^*)$ such that

$$\begin{aligned} &\sum_{i \in \mathbf{N}} (x_i^* - x_i) u_i(\mathbf{x}^*) \\ &+ \sum_{i,j \in \mathbf{N},\ i \neq j} (h_{ij} - h_{ij}^*) c_{ij}(\mathbf{h}^*) \geq 0 \ \forall (\mathbf{x}, \mathbf{h}) \in H. \end{aligned} \tag{6.12}$$

This equivalence result follows from the fact that (6.10), (6.11) represent an analogue of *Karush-Kuhn-Tucker* conditions for variational inequality (6.12); see Sections 11.1 and 12.5.

Exercise 6.3. By using the results of Section 11.2, find existence conditions for this model.

Further analysis of the problem relies upon the properties of the mappings u and c. It is possible to utilize the results of the general theory for investigating various kinds of applications.

Part II

COMPLEMENTARITY PROBLEMS

The complementarity problems (CP's for short) can be viewed as a subclass of the variational inequalities (VI's for short) where the feasible set is a cone or a cone segment. These essential features of CP's enable one to develop very effective tools for their investigation and solution. More precisely, many existence and uniqueness results and solution methods for CP's were established and substantiated under essentially weaker assumptions than those for general VI's. At the same time, various classes of equilibrium problems, which are formulated as CP's, were described in Part I. In this part, we present some of basic results in this field and discuss their applications to the models described in the previous part.

Chapter 7

Complementarity with Z Properties

This chapter is devoted to the general theory of CP's and to its specializations for problems whose cost mappings possess so-called Z or off-diagonal antitone properties. These problems also admit very effective solution methods.

7.1 Classes of complementarity problems

Let X be a nonempty, closed and convex cone in a finite - dimensional Euclidean space E, $G : X \to E$ a continuous mapping. Denote by $X' = \{z \in E \mid x^T z \geq 0 \quad \forall x \in X\}$ the dual (conjugate) cone to X. Then (see (1.16)) we can define the general CP as the problem of finding a point $x^* \in E$ such that

$$x^* \in X, G(x^*) \in X', (x^*)^T G(x^*) = 0. \tag{7.1}$$

This problem can be equivalently rewritten as VI with the cost mapping G and the feasible set X, as the following proposition states.

Proposition 7.1. *CP (7.1) is equivalent to VI: Find a point $x^* \in X$ such that*

$$(x - x^*)^T G(x^*) \geq 0 \quad \forall x \in X. \tag{7.2}$$

Proof. If x^* solves problem (7.1), then for each $x \in X$, we have

$$(x - x^*)^T G(x^*) = x^T G(x^*) \geq 0$$

since $G(x^*) \in X'$. Conversely, let x^* solve problem (7.2). If we suppose that $(x^*)^T G(x^*) > 0$, then letting $x = 0$ in (7.2) gives the contradiction

$$0 > -(x^*)^T G(x^*) = (0 - x^*)^T G(x^*) \geq 0,$$

hence $(x^*)^T G(x^*) \leq 0$. If $(x^*)^T G(x^*) < 0$, then letting $x = \alpha x^* \in X$ with $\alpha > 1$ in (7.2) again gives the contradiction

$$0 > (\alpha - 1)(x^*)^T G(x^*) = (x - x^*)^T G(x^*) \geq 0.$$

It follows that $(x^*)^T G(x^*) = 0$, hence

$$(x - x^*)^T G(x^*) = x^T G(x^*) \geq 0$$

for each $x \in X$, i.e. $G(x^*) \in X'$ and x^* solves CP (7.1). □

Among various classes of CP's, the following ones are most investigated. The *standard complementarity problem* corresponds to the case where $E = \mathbb{R}^n$ and X is the non-negative orthant $\mathbb{R}^n_+ = \{x \in \mathbb{R}^n \mid x_i \geq 0 \quad \forall i = 1, \dots, n\}$ in $\mathbb{R}^n$. Then CP(7.1) can be rewritten in the standard form:

$$x_i^* \geq 0, \; G_i(x^*) \geq 0, \; x_i^* G_i(x^*) = 0 \quad \forall i = 1, \dots, n. \tag{7.3}$$

Exercise 7.1. Prove that problem (7.3) is equivalent to CP (7.1) with $X = \mathbb{R}^n_+$.

The linear complementarity problem (LCP for short) corresponds to the case when G is affine, i.e. $G(x) = Ax + b$, where A is an $n \times n$ matrix, b is a fixed element in $\mathbb{R}^n$.

The standard problem (7.3) involves only the nonnegativity constraints for all the variables whereas many applied CPs contain also either some unrestricted variables or even two-side box constraints. There exists a very suitable and common format for such problems. Set

$$X = \prod_{i=1}^{n} [\alpha_i, \beta_i], \; -\infty \leq \alpha_i < \beta_i \leq +\infty \; \text{ for } \; i = 1, \dots, n; \tag{7.4}$$

and consider the box-constrained VI (7.2), (7.4). This problem is also called the *mixed complementarity problem* (MCP for short). Clearly, it becomes equivalent to CP (7.3) if $\alpha_i = 0$ and $\beta_i = +\infty$ for each $i = 1, \dots, n$. At the same time, bearing in mind Proposition 7.1 and the formulation (7.3), we can write it in more suitable forms.

Proposition 7.2. *VI (7.2), (7.4) is equivalent to each of the following problems:*

(i) Find $x^* \in X$ from (7.4) such that

$$G_i(x^*)(x_i - x_i^*) \geq 0 \quad \forall x_i \in [\alpha_i, \beta_i] \quad \forall i = 1, \ldots, n;$$

and

(ii) Find $x^* \in X$ from (7.4) such that

$$G_i(x^*) \begin{cases} \geq 0 & \text{if } x_i^* = \alpha_i, \\ = 0 & \text{if } x_i^* \in (\alpha_i, \beta_i), \\ \leq 0 & \text{if } x_i^* = \beta_i, \end{cases}$$

for $i = 1, \ldots, n$.

Exercise 7.2. Prove the assertions of Proposition 7.2.

Now we turn to the general existence results for CP's. To this end, we give some results from Chapter 11, which are established for general VI's (Theorems 11.2 and 11.3).

Theorem 7.1. *Let X be a nonempty, convex and closed subset of a finite - dimensional Euclidean space E and let $G : X \to E$ be a continuous mapping. Suppose that there exists a nonempty bounded subset Y of X such that for every $x \in X \backslash Y$ there is $y \in Y$ with*

$$(x - y)^T G(x) > 0. \tag{7.5}$$

Then VI (7.2) is solvable.

Observe that we can set $X = Y$ if X is bounded, but the corresponding result can be used only for bounded MCP's.

Corollary 7.1. *Suppose that $-\infty < \alpha_i < \beta_i < +\infty$ for each $i = 1, \ldots, n$ and that $G : X \to E$ is continuous. Then MCP (7.2), (7.4) has a solution.*

Coercivity condition (7.5) allows us to apply the assertion of Theorem 7.1 to all kinds of CP's, which usually involve unbounded feasible sets. In this case, applying certain additional properties of the cost mapping G and the feasible set, we can transform it into a more suitable format.

7.2 Classes of square matrices and their properties

In this section, we consider some classes of matrices whose properties are strongly related to the corresponding LCP's, moreover, they appear to be very useful for better understanding of properties of nonlinear CP's, which will be presented in this part. An interested reader can find more details in *Cottle, Pang and Stone* (1992), *Ortega and Rheinboldt* (1970), and *Nikaido* (1968); see also the references therein.

Definition 7.1. Let A be an $n \times n$ matrix. The matrix A is said to be
(a) *a Z-matrix* if it has non-positive off-diagonal entries;
(b) *a P-matrix* if all the principal minors of A (i.e. the deferminants of submatrices of A whose entries lie in k selected rows and columns of A for a given k) are positive;
(c) *a P_0-matrix* if all the principal minors of A are nonnegative;
(d) *positive definite* if $x^T Ax > 0$ for each nonzero vector $x \in \mathbb{R}^n$;
(e) *positive semidefinite* if $x^T Ax \geq 0$ for each $x \in \mathbb{R}^n$;
(f) *an M-matrix* if it is a P- and Z-matrix;
(g) *an M_0-matrix* if it is a P_0- and Z-matrix.

From the definitions we obtain the following obvious implications

$$(g) \Longrightarrow (c), (g) \Longrightarrow (a), (f) \Longrightarrow (b), (f) \Longrightarrow (a),$$

but the reverse assertions are not true. Moreover, it will be shown that $(e) \Longrightarrow (c)$ and $(d) \Longrightarrow (b)$, but the reverse assertions are also not true in general.

Exercise 7.3. Give examples of P-matrices (respectively, P_0-matrices) which are not positive definite (respectively, positively semidefinite).

We now give another characterization of P-matrices; see *Cottle, Pang, and Stone* (1992), Theorem 3.3.4.

Proposition 7.3. *Let A be an $n \times n$ matrix. Then the following assertions are equivalent:*
(i) A is a P-matrix,
(ii) A reverses the sign of no nonzero vector, i.e.

$$\sum_{j=1}^{n} a_{ij} x_i x_j \leq 0 \quad \text{for all} \quad i = 1, \ldots, n \Longrightarrow x = 0;$$

(iii) all real eigenvalues of A are positive.

Proof. Clearly, (i) $\Rightarrow$ (ii) for $n = 1$. Using induction, we suppose that this implication holds for $n - 1$ with $n > 1$ and that the $n \times n$ P-matrix A reverses the sign of a nonzero vector $z \in \mathbb{R}^n$. If $z_i = 0$ for some i, then there exists a principal submatrix $\tilde{A}$ of A, which is also a P-matrix, but reverses the sign of a subvector $\tilde{z}$ of z. Since this is a contradiction, no component of z is zero. Set

$$d_i = \frac{1}{z_i} \sum_{j=1}^{n} a_{ij} z_j \leq 0 \quad \text{for} \quad i = 1, \ldots, n$$

and define the diagonal matrix D with the diagonal entries $d_1, \ldots, d_n$. It follows that $(A - D)z = 0$, but the matrix $A - D$ has to be nonsingular, a contradiction, Hence, (i) $\Rightarrow$ (ii).

Suppose that (ii) holds and choose an arbitrary real eigenvalue λ of A and an associate eigenvector z. The vector z must be nonzero and, as λ is real, we can take z to be real. By definition, $Az = \lambda z$, but A does not reverse the sign of z, hence $\lambda > 0$, and (ii) $\Rightarrow$ (iii).

Next, let (iii) hold. Recall that the determinant of A is equal to the product of all its eigenvalues, but the complex eigenvalues of real matrices appear in conjugate pairs. It follows that (i) holds, and the proof is complete. □

Similarly, one can establish another characterization of P_0-matrices; see *Cottle, Pang, and Stone* (1992), Theorem 3.4.2.

Proposition 7.4. *Let A be an $n \times n$ matrix. Then the following assertions are equivalent:*

(i) A is a P_0-matrix,

(ii) for each $x \neq 0$, there exists an index k such that $x_k \neq 0$ and

$$\sum_{j=1}^{n} a_{kj} x_k x_j \geq 0;$$

(iii) all real eigenvalues of A are non-negative.

Exercise 7.4. Following the proof of Proposition 7.3, establish the result of Proposition 7.4.

Taking into account part (ii) of Propositions 7.3 and 7.4, we obtain the relationships between $P(P_0)$- and positive (semi)definite matrices.

Corollary 7.2. *If A is an $n \times n$ positive definite (respectively, positive semidefinite) matrix, then it is a P-matrix (respectively, P_0-matrix).*

Observe that the sum of $P_0(P)$-matrices, unlike that of positive (semi) definite ones, is not a $P_0(P)$-matrix in general.

Example 7.1. The matrices

$$A = \begin{pmatrix} 1 & 5 \\ 0 & 2 \end{pmatrix}$$

and

$$B = \begin{pmatrix} 2 & 0 \\ 5 & 1 \end{pmatrix}$$

are clear P-matrices, but they are not positive semidefinite. Their sum

$$C = A + B = \begin{pmatrix} 3 & 5 \\ 5 & 3 \end{pmatrix}$$

is not a P- (or P_0-) matrix.

We recall that even the smallest perturbations of a positive semidefinite matrix with the help of the unit matrix convert it into a positive definite matrix. We now show that the same property holds true for P_0- and P-matrices.

Proposition 7.5. *An $n \times n$ matrix A is a P_0-matrix if and only if $A + \varepsilon I$ is a P-matrix for each $\varepsilon > 0$.*

Proof. If A is a P_0-matrix, then, by Proposition 7.4, for each $x \neq 0$, there exists an index k such that $x_k \neq 0$ and

$$\sum_{j=1}^{n} a_{kj} x_k x_j \geq 0,$$

hence

$$\sum_{j=1}^{n} a_{kj} x_k x_j + \varepsilon x_k^2 > 0$$

if $\varepsilon > 0$. It means that $A + \varepsilon I$ reverses the sign of no nonzero vector and, by Proposition 7.3, it is a P-matrix. Conversely, fix $x \neq 0$. Since $A + \varepsilon I$ is a P-matrix for each $\varepsilon > 0$, there exists an index i, which may depend on ε, such that

$$\sum_{j=1}^{n} a_{ij} x_j x_i + \varepsilon x_i^2 > 0.$$

Take a sequence $\{\varepsilon_k\} \searrow 0$, then there exists an index s such that

$$\sum_{j=1}^{n} a_{sj} x_j x_s + \varepsilon_{k_l} x_s^2 > 0$$

if $\varepsilon_{k_l} > 0$ for an infinite subsequence $\{k_l\}$. Since $x_s \neq 0$, taking the limit $k_l \to +\infty$ in this inequality yields

$$\sum_{j=1}^{n} a_{sj} x_j x_s \geq 0,$$

therefore, A is a P_0-matrix. □

Thus, P_0- and P-matrices maintain some basic properties of positive semidefinite and positive definite matrices. We now give an additional property of these matrices which is traditionally used in existence theory for LCP's.

Proposition 7.6. *If an $n \times n$ matrix A is a P-matrix (respectively, P_0-matrix), then there exists a vector $x \in \mathbb{R}^n$ such that*

$$x > 0, Ax > 0 \tag{7.6}$$

(respectively, $x \geq 0, x \neq 0, Ax \geq 0$).

Proof. If A is a P-matrix, but (7.6) does not hold, then, using Lemma 3.1 of the alternative, we see that there exists a vector $z \neq 0$ such that $A^T z \leq 0$ and $z \geq 0$. It means that A^T reverses the sign of z, i.e., by Proposition 7.3, A^T is not a P-matrix. Then A is not also a P-matrix.

Next, if A is a P_0-matrix, then, by Proposition 7.5, $A + \varepsilon I$ is a P-matrix for each $\varepsilon > 0$. Choose a sequence $\{\varepsilon_k\} \searrow 0$, then, for each ε_k, there exists $x^k > 0$ such that $Ax^k > -\varepsilon_k x^k$ and, without loss of generality we can suppose that $\|x^k\| = 1$. Then there exists a subsequence $\{x^{k_s}\}$ converging to a point $x \geq 0, x \neq 0$. Taking the corresponding limit, we obtain $Ax \geq 0$. □

Observe that the matrix

$$A = \begin{pmatrix} 1 & 2 \\ 2 & 1 \end{pmatrix}$$

satisfies (7.6) with $x = (1, 1)^T$, but it is not a P_0-matrix, so that the reverse assertions are not true.

Now we turn to the consideration of M-matrices. Being based on the previous results, we can establish an interesting property of their inverses.

Proposition 7.7. *Let A be an arbitrary $n \times n$ Z-matrix. Then A is an M-matrix if and only if the inverse matrix A^{-1} exists and contains only nonnegative entries.*

Proof. If A is an arbitrary $n \times n$ Z-matrix, then it can be represented as follows

$$A = \rho I - B,$$

where B is an $n \times n$ matrix with non-negative entries, $\rho \geq 1$ is some number. Set $\tilde{A} = I - \rho^{-1} B$. Then, by Theorem 2.2 (Hawkins-Simon), $\tilde{A}^{-1}$ exists and is nonnegative if and only if all the principal minors of $\tilde{A}$ are non-negative. However, minors of A and $\tilde{A}$ have the same signs. It follows that A^{-1} exists and nonnegative if and only if all the principal minors of A are nonnegative. □

We can now conclude that the problem

$$Ax = b, x \geq 0,$$

where A is an M-matrix and b is a non-negative vector, is equivalent to the standard system of linear equations

$$Ax = b,$$

which admits the explicit form of its solution. Note that this property was used in the analysis of the open input-output model; see Section 2.1. In other words, the "non-classical" part of the problem becomes superfluous if A is an M-matrix, and this formulation of various applied problems is very popular; see e.g. *Ortega and Rheinboldt* (1970). Moreover, this approach of the tacit elimination of "non-classical" constraints may be extended to some general nonlinear problems and is developed in degree theory. At the same time, it was noticed in Section 2.2, that even small modifications of the model create serious difficulties for this approach and force us to study more "stable" models of equilibrium, such as CP's.

7.3 Complementarity problems with Z cost mappings

The concept of a Z-mapping is a direct extension of that of a Z-matrix. Denote by e^j the j-th coordinate vector in $\mathbb{R}^n$, i.e.

$$e_i^j = \begin{cases} 0 & \text{if} \quad i \neq j, \\ 1 & \text{if} \quad i = j. \end{cases}$$

Definition 7.2. A mapping $G : X \to \mathbb{R}^n$ is said to be

(a) *a Z-mapping*, if, for each $x \in X$, the function $\mu_{ij}(\tau) = G_i(x + \tau e^j)$ is nonincreasing for all $i \neq j$;

(b) *off-diagonally antitone*, if, for each pair of points $x', x'' \in X$ such that $x' \geq x''$, it holds that $G_k(x') \leq G_k(x'')$ for each k such that $x'_k = x''_k$.

Although the classes of Z-mappings and off-diagonally antitone mappings seem somewhat different, they in fact coincide as the following proposition states.

Proposition 7.8. *The classes of Z- and off-diagonally antitone mappings coincide.*

Proof. Suppose that $G : X \to \mathbb{R}^n$ is a Z-mapping. Choose an arbitrary pair of points $x', x'' \in X$, such that $x' \geq x''$, then $x' = x'' + \sum_{j=1}^{n} \tau_j e^j$ where

$\tau_j \geq 0$. It follows that $G_i(x') = G_i(x'' + \sum_{j\neq i} \tau_j e^j) \leq G_i(x'')$ for each index i such that $x'_i = x''_i$, i.e. G is off-diagonally antitone. Conversely, if G is off-diagonally antitone and $x' = x'' + \tau e^j$ for fixed j with $\tau > 0$, then $x' \geq x''$. Taking any index $i \neq j$, we obtain $x'_i = x''_i$ and $G_i(x') \leq G_i(x'')$, i.e. G is a Z-mapping. □

Thus, Proposition 7.8 presents another characterization of Z-mappings. From the definition we obtain some other characterization in the differentiable case. Then there exists the Jacobian of G at x, denoted by $\nabla G(x)$, whose entries are partial derivatives $\frac{\partial G_i(x)}{\partial x_j}$ for $i, j = 1, \ldots, n$.

Proposition 7.9. *A differentiable mapping $G : \mathbb{R}^n \to \mathbb{R}^n$ is a Z-mapping if and only if its Jacobian ∇G is a Z-matrix at each point $x \in \mathbb{R}^n$.*

Exercise 7.5. Prove the assertion of Proposition 7.9.

If G is an affine mapping, i.e. $G(x) = Ax + b$, then, by Proposition 7.9, G is a Z-mapping if and only if A is a Z-matrix.

We shall consider the standard CP (7.3) in the case when G is a Z-mapping. The essential features of such problems enable us to present special tools for their investigation and solution finding. First we define the auxiliary (or otherwise, feasible) set of CP (7.3) as follows:

$$D = \{x \in \mathbb{R}^n \mid x \geq 0, G(x) \geq 0\}.$$

CP is said to be *feasible* if D is nonempty.

It appears D possesses certain vector lattice properties. For each pair of points $x, y \in \mathbb{R}^n$, we can define their component-wise minimal point (meet) $z = \min\{x, y\}$ as follows:

$$z_i = \min\{x_i, y_i\} \quad \text{for} \quad i = 1, \ldots, n.$$

Lemma 7.1. *If $G : \mathbb{R}^n_+ \to \mathbb{R}^n$ is a Z-mapping, then the set D contains $\min\{x, y\}$ for arbitrary points $x, y \in D$.*

Proof. Fix $x, y \in D$ and set $z = \min\{x, y\}$. Then clearly $z \geq 0$. Define the index sets $I_1 = \{i \mid x_i < y_i\}$ and $I_2 = \{i \mid x_i \geq y_i\}$. If $I_1 = \emptyset$, then $z = y \in D$ and similarly if $I_2 = \emptyset$, then $z = x \in D$. Let us consider the case when $I_1 \neq \emptyset$ and $I_2 \neq \emptyset$. If $i \in I_1$, then $z_i = x_i$, but $z \leq x$ and, in view of Proposition 7.8, $G_i(z) \geq G_i(x) \geq 0$. Similarly, if $i \in I_2$, then $z_i = y_i$, but $z \leq y$ and $G_i(z) \geq G_i(y) \geq 0$. Therefore, $G(z) \geq 0$ and $z \in D$. □

For the set D, we can define its minimal element

$$\min D = \{z \in D \mid z \leq x \quad \forall x \in D\}.$$

From the definition it follows that $\min D$ is unique.

Lemma 7.2. *If the set D is nonempty, $G : \mathbb{R}^n_+ \to \mathbb{R}^n$ is a continuous Z-mapping, then the optimization problem*

$$\min_{x \in D} \to \sum_{i=1}^{n} x_i \tag{7.7}$$

has a solution, which coincides with $\min D$.

Proof. Since D is nonempty, take an arbitrary point $y \in D$. Set $\mu(x) = \sum_{i=1}^{n} x_i$ and for an arbitrary number $\gamma > \mu(y)$ define the set

$$D_\gamma = D \bigcap \{x \in \mathbb{R}^n \mid \mu(x) \leq \gamma\}.$$

Clearly, D is bounded and also closed since G is continuous. Due to the *Weierstrass* theorem, the function μ attains its minimal value on D_γ, hence, on D as well. Thus, problem (7.7) has a solution, say u. If $u \neq \min D$, then there exist a point $v \in D$ and at least one index i such that $v_i < u_i$. Set $z = \min\{u, v\}$, then $z \in D$ on account of Lemma 7.1, but $\mu(z) < \mu(u)$, a contradiction. Therefore, $\min D$ exists and is a (unique) solution to (7.7). □

We are now ready to establish an existence result for CP (7.3).

Theorem 7.2. *If the set D is nonempty, $G : \mathbb{R}^n_+ \to \mathbb{R}^n$ is a continuous Z-mapping, then CP (7.3) has a solution, which coincides with* $\min D$.

Proof. From Lemma 7.2 we have that the element $z = \min D$ exists. Suppose that there exists an index k such that

$$z_k > 0 \quad \text{and} \quad G_k(z) > 0.$$

Then, due to the continuity of G, we have

$$z_k - \varepsilon \geq 0 \quad \text{and} \quad G_k(z - \varepsilon e^k) \geq 0$$

for $\varepsilon > 0$ sufficiently small. However, $z \geq z - \varepsilon e^k$ and

$$G_i(z - \varepsilon e^k) \geq G_i(z) \geq 0 \quad \text{for each} \quad i \neq k,$$

hence $z - \varepsilon e^k \in D$, which contradicts the definition of z. Therefore

$$z_i G_i(z) = 0 \quad \text{for all} \quad i = 1, \dots, n,$$

and z solves CP (7.3). □

Figure 7.1:

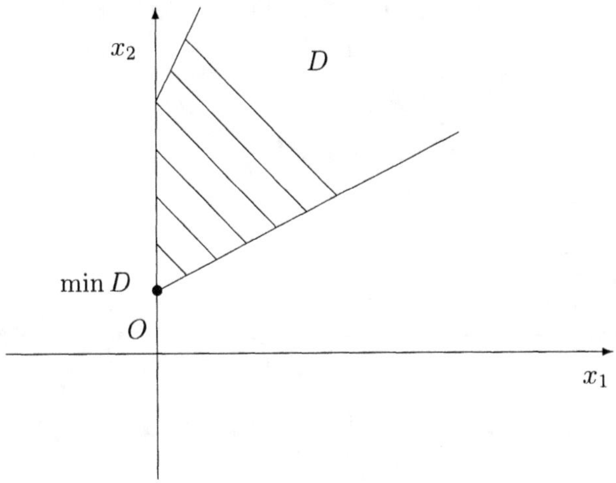

Observe that CP (7.3) may have many solutions even under the conditions of Theorem 7.2, as the following example illustrates.

Example 7.2. Let us consider the following LCP:

$$x^* \geq 0, Ax^* + b \geq 0, (x^*)^T(Ax^* + b) = 0,$$

where

$$x^* = (x_1^*, x_2^*)^T, A = \begin{pmatrix} a_{11} & a_{12} \\ a_{21} & a_{22} \end{pmatrix}$$

with $a_{12} \leq 0$, $a_{21} \leq 0$, and $b = (b_1, b_2)^T$.

(a) If we set

$$A = \begin{pmatrix} 2 & -1 \\ -1 & 2 \end{pmatrix}, b = (4, -2)^T,$$

when

$$D = \{x \geq 0 \mid 2x_1 - x_2 + 4 \geq 0, -x_1 + 2x_2 - 2 \geq 0\}$$

and

$$\min D = (0, 1)^T.$$

This LCP has clearly the unique solution $\min D$; see Figure 7.1.

Figure 7.2:

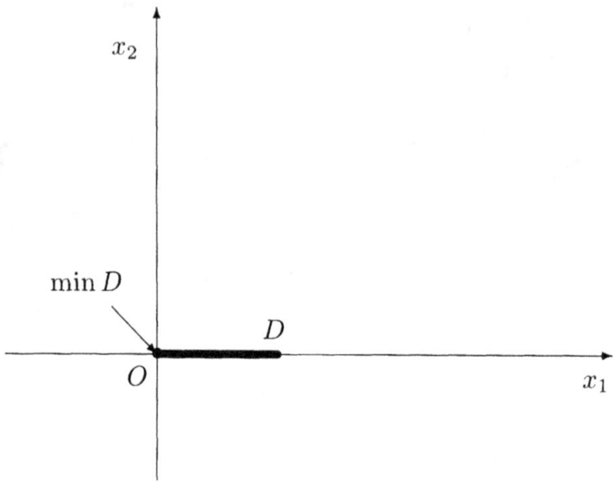

(b) If we set

$$A = \begin{pmatrix} 0 & -1 \\ -1 & 0 \end{pmatrix}, b = (0, 2)^T,$$

when

$$D = [0, 2] \times \{0\},$$

i.e. it coincides with the solution set and clearly

$$\min D = (0, 0)^T \in D.$$

This LCP has clearly the infinite number of solutions; see Figure 7.2.

Thus, Z properties allow us to obtain some other existence result of solutions, which is based on the feasibility of the initial problem. In order to ensure feasibility, we can make use of some other properties of the cost mapping.

Exercise 7.6. Derive from Proposition 7.6 the feasibility of LCP (7.3) with $G(x) = Ax + b$, where A is a P-matrix.

It follows that such a LCP with A being an M-matrix always has a solution, moreover, this solution is the minimal element of its feasible set.

At the same time, there exists a different way of proving the result of Theorem 7.2 which is based upon constructing an iterative sequence.

For the sake of brevity, set

$$(x_{-i}, \alpha) = (x_i, \dots, x_{i-1}, \alpha, x_{i+1}, \dots, x_n)^T$$

for a point $x = (x_1, \dots, x_n)^T$, a number α, and an index i. Let us consider the following Jacobi type iterative algorithm.

Algorithm (Jacobi). Choose a point $x^0 \in D$. At the k-th iteration, $k = 0, 1, \dots,$ we have a point x^k and compute the next iterate x^{k+1} componentwise. Namely, for each index $i = 1, \dots, n$, we set $x_i^{k+1} = 0$ if $G_i(x_{-i}^k, 0) \geq 0$, otherwise we set $x_i^{k+1}$to be the number in $(0, x_i^k]$ such that $G_i(x_{-i}^k, x_i^{k+1}) = 0$.

Theorem 7.3. *If the set D is nonempty, $G : \mathbb{R}_+^n \to \mathbb{R}^n$ is a continuous Z-mapping, then the above algorithm is well-defined and generates a sequence $\{x^k\}$ converging to a solution of CP (7.3).*

Proof. Since $D \neq \emptyset$, we can choose a starting point x^0 in D. We now proceed to show that

$$\begin{array}{c} 0 \leq x^{k+1} \leq x^k, x^k \in D, x_i^{k+1} G_i(x_{-i}^k, x_i^{k+1}) = 0 \\ \forall i = 1, \dots, n, \end{array} \tag{7.8}$$

for $k = 0, 1, \dots$ In fact, the first and third relations in (7.8) are fulfilled by construction. Next, $x^{k+1} \leq (x_{-i}^k, x_i^{k+1})$, but G is a Z-mapping, hence $G_i(x^{k+1}) \geq G_i(x_{-i}^k, x_i^{k+1}) \geq 0$, i.e. $x^{k+1} \in D$ if $x^k \in D$. Moreover, if $G_i(x^k) \geq 0$ and $G_i(x_{-i}^k, 0) < 0$, then, by continuity of G, there exists a number $x_i^{k+1} \in (0, x_i^k)$ such that $G_i(x_{-i}^k, x_i^{k+1}) = 0$. It means that the algorithm is well-defined, but the sequence $\{x^k\}$ must converge to a point x^* in view of the first relation in (7.8). Taking the limits in the other relations in (7.8) gives

$$x^* \in D, x_i^* G_i(x^*) = 0 \quad \text{for} \quad i = 1, \dots, n;$$

hence x^* is a solution to CP (7.3). □

We can now deduce the assertion of Theorem 7.2 from Theorem 7.3 and Lemma 7.2.

Corollary 7.3. *If the set D is nonempty, $G : \mathbb{R}_+^n \to \mathbb{R}^n$ is a continuous Z-mapping, then the minimal element* $\min D$ *exists and solves CP (7.3).*

Proof. From Lemma 7.2 it follows that $\min D$ exists, hence we can take $x^0 = \min D$ in the above algorithm. Then the sequence $\{x^k\}$ converges to a solution x^* of CP (7.3) due to Theorem 7.3, but, on account of (7.8), we have

$$x^* \leq \min D, \quad \text{hence} \quad x^* = \min D,$$

since $\min D$ is determined uniquely. □

Thus, the Jacobi algorithm can be applied to finding solutions of feasible CP's with Z-mappings. Moreover, the results of this section can be extended to MCP (7.2), (7.4) with the corresponding modifications of the auxiliary set D and the Jacobi algorithm.

Exercise 7.7. Define

$$D = \{x \in X \mid x_i < \beta_i \Rightarrow G_i(x) \geq 0 \quad \forall i = 1, \dots, n\}$$

for MCP (7.2), (7.4) and show that the assertion of Theorem 7.2 remains true if $\alpha_i > -\infty$ for $i = 1, \dots, n$ and $\mathbb{R}^n_+$ is replaced with X. Make the modification of the Jacobi algorithm which allows for obtaining the result of Theorem 7.3.

Together with the Jacobi algorithm one can apply also similar extensions of other known algorithms of solving systems of equations.

Chapter 8

Applications

In the chapter, we consider applications of CP's with Z - mappings, which were presented in Chapter 7, to input-output and price equilibrium models.

8.1 Input-output models

First we consider the open input-output model described in Section 2.1. Recall that the classical formulation of this model (see (2.2)) corresponds to the system:

$$(I - A)x^* = y, \quad x^* \geq 0, \tag{8.1}$$

where I is the $n \times n$ unit matrix, A is the $n \times n$ input-output matrix with entries $a_{ij}, y = (y_1, \dots, y_n)^T$ is a given final demand vector, and $x^* = (x_1^*, \dots, x_n^*)^T$ is a vector of unknown outputs. In Section 2.1, several sufficient conditions, which are based on the Perron-Frobenius theorems for nonnegative matrices, were presented; see Theorems 2.2, 2.3, and 2.5. They ensure for system (8.1) to be consistent. However, after a suitable extension of this system, one can guarantee existence of a solution under more general conditions. Namely, let us consider the following LCP: Find $x^* \in \mathbb{R}^n$ such that

$$x^* \geq 0, (I - A)x^* - y \geq 0, (x^*)^T[(I - A)x^* - y] = 0; \tag{8.2}$$

which is clearly a particular case of CP (7.3) where

$$G(x) = (I - A)x - y. \tag{8.3}$$

By definition, the input-output coefficients a_{ij} are nonnegative, hence the Jacobian $\nabla G(x) = I - A$ is a Z-matrix, i.e., by Proposition 7.9, G is now a Z-mapping. Therefore, we can apply the results of Chapter 7 to establish existence of solutions of problem (8.2). In fact, the following assertion is a direct consequence of Theorem 7.2.

Theorem 8.1. *Suppose that the set*

$$D = \{x \in \mathbb{R}^n \mid x \geq 0, x - Ax - y \geq 0\} \tag{8.4}$$

is nonempty. Then LCP (8.2) has a solution, which coincides with $\min D$.

In comparison with system (8.1), problem (8.2) provides the equality

$$x_i^* - \sum_{j=1}^{n} a_{ij} x_j^* = y_i$$

only for commodities with positive outputs, i.e. in case $x_i^* > 0$. This approach becomes much more suitable in the presence of two-side constraints (see (2.2′)), which seem very natural for input-output problems. However, the Perron-Frobenius theory then does not guarantee the consistency of the system (2.2′), but we can replace it by the MCP: Find $x^* \in X$ such

$$(x - x^*)^T G(x^*) \geq 0 \quad \forall x \in X, \tag{8.5}$$

where $X = [x', x'']$ and G is defined in (8.3) (see also Proposition 7.2 for related formulations of this problem). Then, using Exercise 7.7, we can obtain similar existence results for problem (8.5) as well.

Exercise 8.1. Replace D in (8.4) by

$$D = \left\{ x \in X \,\middle|\, x_i < x_i'' \Longrightarrow x_i - \sum_{j=1}^{n} a_{ij} x_j - y_i \geq 0 \text{ for } \quad i = 1, \dots, n \right\}$$

and show that the assertion of Theorem 8.1 remains true for problem (8.5).

Moreover, Theorem 7.3 justifies the application of the Jacobi algorithm to find a solution of problem (8.2), which can be also adjusted for (8.5).

Algorithm (Jacobi). Choose a point $x^0 \in D$, where D is defined in (8.4). At the k-th iteration, $k = 0, 1, \dots$, we have a point x^k and compute the next iterate x^{k+1} componentwise. Namely, for each index $i = 1, \dots, n$, we set $x_i^{k+1} = 0$ if

$$\sum_{j \neq i} a_{ij} x_j^k + y_i \leq 0,$$

otherwise we set x_i^{k+1} to be the number in $(0, x_i^k]$ such that

$$x_i^{k+1} = a_{ii} x_i^{k+1} + \sum_{j \neq i} a_{ij} x_j^k + y_i.$$

The above computational procedure is very simple and does not differ essentially from the well-known Jacobi algorithm for systems of linear equations. The economical sense of this algorithm is also very clear. Being based on the current output vector x^k, each branch of the economy computes its own output separately for finding equilibrium in its commodity, which coincides with the balance equation if the output is positive. The reactions of all the branches constitute the next iterate x^{k+1} and so on. Theorem 7.3 ensures the convergence of this natural process.

Theorem 8.2. *If the set D in (8.4) is nonempty, the above Jacobi algorithm is well-defined and generates a sequence $\{x^k\}$ converging to a solution of LCP (8.2).*

Similarly, taking into account Exercises 7.7 and 8.1, we can adjust the Jacobi algorithm to MCP (8.3), (8.5).

Exercise 8.2. Describe the modification of the Jacobi algorithm for MCP (8.3), (8.5) and prove its convergence.

The classical input-output model is based upon the linear technology assumptions, i.e. the industrial demand Ax depends linearly on the output vector x. However, some technologies may be nonlinear, then we have to replace the map $x \mapsto Ax$ with $x \mapsto A(x)$. Let us consider the following property of industrial demand.

Definition 8.1. A mapping $A : X \to \mathbb{R}^n$ is said to be *isotone*, if, for each pair of points $x', x'' \in X$ such that $x' \geq x''$, it holds that $A(x') \geq A(x'')$.

Since the input-output coefficients a_{ij} are assumed to be non-negative, the mapping $x \mapsto Ax$ is clearly isotone, and this property seems very natural even for a nonlinear industrial demand mapping $A(x)$. At the same time, this is sufficient for applying the results of Section 7, as the following proposition states.

Proposition 8.1. *If $A : X \to \mathbb{R}^n$ is an isotone mapping, then the mapping $B : X \to \mathbb{R}^n$, defined by*

$$B(x) = x - A(x),$$

is a Z-mapping.

Proof. Take an arbitrary pair of points $x', x'' \in X$ such that $x' \geq x''$ and choose any index k such that $x'_k = x''_k$. Since A is isotone, we have

$$\begin{aligned} B_k(x') - B_k(x'') &= [x'_k - A_k(x')] - [x''_k - A_k(x'')] \\ &= A_k(x'') - A_k(x') \leq 0, \end{aligned}$$

i.e. B is off-diagonally antitone. By Proposition 7.8, it means that B is a Z-mapping. □

Therefore, if the industrial demand is represented by a nonlinear isotone mapping $A : \mathbb{R}^n_+ \to \mathbb{R}^n$, we can consider the following extension of model (8.2) in the form of CP: Find $x^* \in \mathbb{R}^n_+$ such that

$$x^* \geq 0, x^* - A(x^*) - y \geq 0, (x^*)^T[x^* - A(x^*) - y] = 0, \tag{8.6}$$

which is also a particular case of CP (7.3) where

$$G(x) = x - A(x) - y \tag{8.7}$$

with G being a Z-mapping due to Proposition 8.1. For this reason, we can apply Theorems 7.2 and 7.3 to obtain existence results and to construct iteration sequences converging to its solution.

Theorem 8.3. *Suppose that* $A : \mathbb{R}^n_+ \to \mathbb{R}^n$ *is a continuous isotone mapping and that the set*

$$D = \{x \in \mathbb{R}^n \mid x \geq 0, x - A(x) - y \geq 0\}$$

is nonempty. Then:

(i) problem (8.6) has a solution, which coincides with $\min D$*;*

(ii) the Jacobi algorithm, described in Section 7.3, is well-defined and generates a sequence $\{x^k\}$ *converging to a solution of problem (8.6).*

Exercise 8.3. Prove the assertions of Theorem 8.3.

Similarly, using the VI formulation (8.5) with G in (8.7), we can give the nonlinear input-output model in the presence of two-side constraints on outputs. The analysis in this case extends the previous results for the case of the linear industrial demand.

8.2 Price equilibrium models

Let us consider the Walrasian price equilibrium model from Section 5.2. This model describes an economy with perfect competition, which deals in n commodities. In the model, the equilibrium price vector $p^* \in \mathbb{R}^n$ satisfies the following conditions (see (5.3)):

$$p^* \geq 0, E(p^*) \leq 0, (p^*)^T E(p^*) = 0, \tag{8.8}$$

where

$$E(p) = D(p) - S(p)$$

describes the excess demand at price p, so that $D(p)$ and $S(p)$ are the values of demand and supply at price p, respectively. Obviously (8.8) is a particular case of CP (7.3) with $G(p) = -E(p)$.

One of the most popular assumptions on demand or excess demand is the gross substitutability; see *Arrow and Hahn* (1971), *Morishima* (1964), and *Nikaido* (1968).

Definition 8.2. Let X be a box constrained set in $\mathbb{R}^n_+$, i.e.

$$X = \prod_{i=1}^{n} [\alpha_i, \beta_i], 0 \leq \alpha_i < \beta_i \leq +\infty \quad i = 1, \ldots, n. \tag{8.9}$$

A mapping $A : X \to \mathbb{R}^n$ is said to be *gross substitutable* (or briefly, a *GS-mapping*), if, for each pair of points $x', x'' \in X$ such that $x' \geq x''$ and $I(x', x'') = \{i \mid x'_i = x''_i\}$ is nonempty, there exists an index $k \in I(x', x'')$ with $A_k(x') \geq A_k(x'')$.

Observe that the gross substitutability of (excess) demand means that any transition to a greater price vector implies the non decrease of (excess) demand for at least one commodity whose price does not change. In other words, each subset of commodities in such a market contains substitutable ones. There exists a simple relationship between GS- and Z-mappings.

Proposition 8.2. *Let $A : X \to \mathbb{R}^n$ be a mapping with X being defined in (8.9). Then:*

(i) if $-A$ is a Z-mapping, then A is a GS-mapping;

(ii) if A is a continuous GS-mapping, then $-A$ is a Z-mapping.

Proof. Assertion (i) follows from the definitions and Proposition 7.8. Conversely, suppose that A is a continuous GS-mapping. Choose arbitrary points $x', x'' \in X$ such that $x' \geq x'', x' \neq x''$ and any index $k \in I(x', x'')$. Define the points y', y'' as follows:

$$y'_i = \begin{cases} x'_i + \varepsilon & \text{if} \quad i \in I(x', x''), i \neq k, x'_i < \beta_i, \\ x'_i & \text{otherwise;} \end{cases}$$

$$y''_i = \begin{cases} x''_i - \varepsilon & \text{if} \quad i \in I(x', x''), i \neq k, x''_i > \alpha_i, \\ x''_i & \text{otherwise;} \end{cases}$$

Then $y', y'' \in X$ for $\varepsilon > 0$ small enough, $y' \geq y''$, but $y'_i > y''_i$ if $i \neq k$. Since A is a GS-mapping, then $A_k(y') \geq A_k(y'')$. Taking the limit $\varepsilon \to 0$, we then obtain $A_k(x') \geq A_k(x'')$ by continuity. Since k was taken arbitrarily, if follows that $-A_i(x') \leq -A_i(x'')$ for each $i \in I(x', x'')$, i.e. $-A$ is a Z-mapping. □

We can give another definition of gross substitutability for differentiable mappings.

Definition 8.3. Let X be defined in (8.9). A differentiable mapping $A : X \to \mathbb{R}^n$ is said to be gross substitutable (or a GS-mapping) if its Jacobian $\nabla A(x)$ has non-negative off-diagonal entries.

Exercise 8.4. Taking into account Proposition 8.2 (see also Proposition 7.9) show that the concept of the GS-mapping in Definition 8.3 is equivalent to that in Definition 8.2 if A is differentiable.

Obviously, the concept from Definition 8.3 is more suitable for verification. Nevertheless, Proposition 8.2 says that we can in principle apply all the results of Section 7.3 to equilibrium problem (8.8) if E is a GS-mapping.

However, some classes of price equilibrium problems possess special properties which may require deeper considerations for obtaining similar results. For instance, the domain of E does not usually include the boundary of $\mathbb{R}^n_+$, i.e. the (excess) demand may be undefined for zero prices. At the same time, they may include the third relation in (8.8) as axiom (*Walras law*), which holds for each price vector $p > 0$. These conditions do not prevent from the application of the Jacobi algorithm, described in Section 7.3, for finding a solution of CP (8.8). In fact, if $E(p)$ is undefined at some point p from the boundary of $\mathbb{R}^n_+$ and there exists a sequence $\{p^k\}$ converging to p, such that $p^k \in \mathbb{R}^n_> = \{p \in \mathbb{R}^n \mid p > 0\}$, then $\lim_{k\to\infty} E_i(p^k) = +\infty$ for some i. This set $\mathbb{R}^n_>$ is called the *positive orthant* and clearly $\mathbb{R}^n_> = \text{int}\mathbb{R}^n_+$. It follows that the feasible set

$$D = \{p \in \mathbb{R}^n \mid p \geq 0, E(p) \leq 0\}$$

does not intersect the boundary of $\mathbb{R}^n_+$ and that E can be supposed to be continuous on D. Therefore, the results of Section 7.3 remain true.

In particular, we see that the k-th iteration of the Jacobi algorithm consists in equilibrating the excess demand for each separate commodity by driving the price for this commodity, considering other current prices as fixed. Theorem 7.3 establishes convergence of this natural price adjustment process to an equilibrium vector.

Moreover, if the equilibrium model involves price rigidities, then the constraint $p \geq 0$ is replaced with $p \in X$, where X is defined in (8.9) and $\alpha_i > 0$ for $i = 1, \dots, n$. Then (8.8) should simply be replaced with (8.5), $G = -E$ is usually continuous on X and, taking into account Exercise 7.7, we can apply both existence results and the Jacobi algorithm to this model.

8.3 A pure trade market model

We now consider a specialization of the previous Walrasian price equilibrium model. This model also describes a market structure dealing in n commodities without production, i.e. it is an exchange or a pure trade model. So, there are m economic agents, who are both consumers and traders. The i-th economic agent possesses the endowments $w^{(i)} = (w_1^{(i)}, \dots, w_n^{(i)})^T \geq 0$ and is supposed to have the Cobb-Douglas utility function

$$u_i(x) = x_1^{\sigma_{i1}} \times \cdots \times x_n^{\sigma_{in}},$$

where $\sigma^{(i)} = (\sigma_{i1}, \dots \sigma_{in})^T \geq 0$ represents the vector of commodities weights, and without loss of generality we suppose that

$$\sum_{j=1}^{n} \sigma_{ij} = 1.$$

Therefore, the supply is fixed, i.e. $S(p) \equiv S$ where the total endowment of the j-th commodity is given by

$$S_j = \sum_{i=1}^{m} w_j^{(i)} \quad \text{for} \quad j = 1, \dots, n.$$

Additionally, we suppose that each economic agent possesses at least one commodity, i.e. $w^{(i)} \neq 0$ for $i = 1, \dots, m$. Then, given a price vector $p = (p_1, \dots, p_n) > 0$, the demand of the i-th agent is determined as the unique solution x^i of the optimization problem:

$$\max \to u_i(x)$$

subject to

$$p^T x \leq p^T w^{(i)}, x \geq 0.$$

Namely, the solution is the following:

$$x_j^i = \frac{1}{p_j} \sigma_{ij} p^T w^{(i)} \quad \text{for} \quad j = 1, \dots, n. \tag{8.10}$$

Exercise 8.5. Prove formula (8.10).

It follows that the market demand for the j-th commodity is given by

$$D_j(p) = \frac{1}{p_j} \sum_{i=1}^{m} \sigma_{ij} p^T w^{(i)}$$

and that the market excess demand for this commodity is given by

$$E_j(p) = \frac{1}{p_j} \sum_{i=1}^{m} \sigma_{ij} p^T w^{(i)} - S_j \quad \text{for} \quad j = 1, \dots, n. \tag{8.11}$$

Although the classical version (8.8), (8.11) of this model was investigated rather intensively for a long period from somewhat different points of view, we now describe this model only to illustrate applications of the concept of the Z-mapping and related results.

First we note that E is defined on $\mathbb{R}^n_>$ and it is differentiable, so that

$$\frac{\partial E_j(p)}{\partial p_l} = \frac{1}{p_j} \sum_{i=1}^{m} \sigma_{ij} w_l^{(i)} \geq 0 \quad \text{if} \quad j \neq l,$$

hence, by Definition 8.3 and Exercise 8.4, E is a GS-mapping. Also, by Proposition 7.9, $-E$ is a Z-mapping. Moreover,

$$\begin{aligned} p^T E(p) &= \sum_{j=1}^{n} p_j E_j(p) = \sum_{j=1}^{n} \sum_{i=1}^{m} \sigma_{ij} p^T w^{(i)} - \sum_{j=1}^{n} p_j S_j \\ &= \sum_{i=1}^{m} p^T w^{(i)} - \sum_{j=1}^{n} p_j S_j = p^T S - p^T S = 0, \end{aligned}$$

i.e. the Walras law holds. It means that feasibility and solvability for the classical model (8.8), (8.11) are equivalent. Nevertheless, the Jacobi algorithm can be applied to find its equilibrium vectors. Of course, we can first consider the system of linear equations

$$\sum_{i=1}^{m} \sigma_{ij} p^T w^{(i)} - p_j S_j = 0 \quad \text{for} \quad j = 1, \dots, n$$

instead of (8.8), (8.11) and find its positive solutions. However, if we consider the slightly modified model with price rigidities, e.g. when $p \in X$ with X being defined in (8.9) and $\alpha_i > 0$ for $i = 1, \dots, n$, then (8.8), (8.11) is replaced with (8.5) and $G = -E$, defined in (8.11). Both the existence results and the Jacobi algorithm become rather useful for this model. These results may be applied to the extended model with additional "single-commodity" consumers and producers, which is described in Section 10.1.

8.4 Price oligopoly models

We now describe applications of the results of Chapter 7 to the imperfectly competitive equilibrium model in the sense of Bertrand described in Section

5.4. In this model, there are n firms supplying a homogeneous product. If each i-th firm announces its price p_i, then its profit is defined by

$$f_i(p) = p_i q_i(p) - h_i[q_i(p)],$$

where $p = (p_1, \dots, p_n)^T$, $q_i : \mathbb{R}^n_+ \to \mathbb{R}^n$ is the market demand function for the i-th firm and $h_i : \mathbb{R}_+ \to \mathbb{R}$ is its industrial cost function, $i = 1, \dots, n$. As indicated in Section 5.4, the equilibrium point is defined as the Nash equilibrium in the n-person non-cooperative game, where the i-th player has the utility function f_i and the strategy set $\mathbb{R}_+$; see (5.22).

We consider a class of functions q_i and h_i such that the above Nash equilibrium problem is replaced by the CP of form (7.3) where

$$G_i(p) = -\frac{\partial f_i(p)}{\partial p_i} \quad \text{for } i = 1, \dots, n$$

and show that G possesses Z properties. Suppose that the cost functions h_i are affine, i.e.

$$h_i(t) = \gamma_i t + \delta_i \quad \text{where} \quad \gamma_i > 0, \delta_i > 0,$$

and that

$$q_i(p) = \alpha_i \left[\eta(p_1, \dots, p_{i-1}, p_{i+1}, \dots, p_n)/p_i\right]^\kappa - \beta_i,$$

where $\alpha_i > 0$, $\beta_i > 0$, $\kappa \in (0, 1]$, $\eta : \mathbb{R}^{n-1} \to \mathbb{R}_+$ is a non-negative differentiable function, which is non-decreasing in each variable. For instance, we can take

$$\eta(p_1, \dots, p_{i-1}, p_{i+1}, \dots, p_n) = \sum_{j \neq i} \mu_j p_j$$

with $\mu_j \geq 0$ for $j \neq i$.

Exercise 8.6. Show that f_i is concave in p_i under the above assumptions.

Since f_i is concave and differentiable in p_i, Corollary 11.2 yields that the Nash equilibrium problem is equivalent to the following CP:

$$p_i^* \geq 0, G_i(p^*) \geq 0, p_i^* G_i(p^*) = 0 \text{ for } i = 1, \dots, n, \tag{8.12}$$

where

$$G_i(p) = \beta_i - \kappa \alpha_i \gamma_i p_i^{-1} \left[\eta(p_{-i})/p_i\right]^\kappa - (1 - \kappa)\alpha_i \left[\eta(p_{-i})/p_i\right]^\kappa$$

for $i = 1, \dots, n$, with $p_{-i} = (p_1, \dots, p_{i-1}, p_{i+1}, \dots, p_n)$. The mapping G is defined and differentiable on $\mathbb{R}^n_>$, moreover, it is easy to see that its

Jacobian $\nabla G(p)$ is a Z-matrix. Therefore, by Proposition 7.9, $G : \mathbb{R}^n_> \rightarrow \mathbb{R}^n_+$ is a Z-mapping and we can utilize the results from Section 7.3 for establishing existence and for computation of its solution. Let us define the auxiliary set of CP (8.12) as follows

$$D = \{p \in \mathbb{R}^n \mid p \geq 0, G(p) \geq 0\}.$$

Although G may be undefined on the boundary of $\mathbb{R}^n_+$, we see that then $G_i(p^k) \rightarrow -\infty$ for any sequence $\{p^k\}$ converging to a point $p' \in \mathbb{R}^n_+$ with $p'_i = 0$. It follows that D is contained in the domain of G. Applying Theorem 7.2, we conclude that CP (8.12) has a solution, which coincides with $\min D$ if there exists a point $\tilde{p} \geq 0$ such that $G(\tilde{p}) \geq 0$. Moreover, due to Theorem 7.3, the Jacobi algorithm with the starting point $\tilde{p}$ will converge to a solution. Observe that its iteration sequence corresponds to the very natural dynamic adjustment process, where, given a current vector p^k, each i-th firm finds the next price p_i^{k+1} as the optimal reaction for p^k, i.e. it maximizes the profit function

$$\alpha_i \left[\eta(p^k_{-i})\right]^{\kappa} p_i^{1-\kappa} - \alpha_i \gamma_i \left[\eta(p^k_{-i})\right]^{\kappa} p_i^{-\kappa}$$

over $\mathbb{R}_+$.

Exercise 8.7. Write down the specialization of the Jacobi algorithm for the described price oligopoly model.

Chapter 9

Complementarity with P Properties

In Chapter 7, several kinds of square matrices were introduced, including Z- and P-matrices. Being based on the concept of a Z-matrix, we introduced its extension for nonlinear mappings, which implied very useful results in theory and solution methods for corresponding CP's. In this chapter, we consider P type mappings, which can be viewed as extensions of such concepts for matrices.

9.1 Existence and uniqueness results

We start our considerations from the introduction of several concepts of order monotone mappings.

Definition 9.1. Let X be a box constrained set in $\mathbb{R}^n$ of form (7.4). A mapping $G : X \to \mathbb{R}^n$ is said to be

(a) a *P_0-mapping*, if for each pair of points $x', x'' \in X$, $x' \neq x''$, there exists an index i such that $x'_i \neq x''_i$ and

$$(x'_i - x''_i)\,[G_i(x') - G_i(x'')] \geq 0;$$

(b) a *P-mapping*, if for each pair of points $x', x'' \in X$, $x' \neq x''$, it holds that

$$\max_{1 \leq i \leq n} (x'_i - x''_i)\,[G_i(x') - G_i(x'')] > 0;$$

(c) a *strict P-mapping*, if there exists $\varepsilon > 0$ such that $G - \varepsilon I$ is a P-mapping;

(d) an *M_0-mapping*, if it is a P_0- and Z-mapping;

(e) an *M-mapping*, if it is a P- and Z-mapping.

From the above definitions we obtain the following obvious implications

$$(e) \Longrightarrow (d) \Longrightarrow (a), (e) \Longrightarrow (b) \Longrightarrow (a), \text{ and } (c) \Longrightarrow (b),$$

but the reverse assertions are not true in general.

We now give some relationships between these concepts and the corresponding classes of matrices. First we consider the affine case when G is of the form

$$G(x) = Ax + b, \tag{9.1}$$

where A is an $n \times n$ matrix, b is a fixed vector in $\mathbb{R}^n$.

Proposition 9.1. *Suppose G is of form (9.1). Then G is a P_0-(respectively, P-) mapping if and only if A is a P_0-(respectively, P-) matrix.*

This assertion follows from Propositions 7.4 and 7.3. Taking into account Example 7.1, we now conclude that the classes of P_0- and P-mappings are not closed with respect to addition.

Exercise 9.1. Taking into account Propositions 7.9 and 9.1, prove that a mapping G of form (9.1) is an M_0-(respectively, M-)mapping if and only if A is an M_0-(respectively, M-) matrix.

We now turn to the differentiable case.

Proposition 9.2. *(Facchinei and Pang, Proposition 3.5.9) Let $G : X \to \mathbb{R}^n$ be a continuously differentiable mapping on a rectangle set $X \subseteq \mathbb{R}^n$. Then the following assertions hold:*

(i) If $\nabla G(x)$ is a P-matrix for all $x \in X$, then G is a P-mapping.

(ii) If $\nabla G(x)$ is a P_0-matrix for all $x \in X$, then G is a P_0-mapping.

(iii) If X is open and G is a P_0-mapping, then $\nabla G(x)$ is a P_0-matrix for all $x \in X$.

Observe that the reverse assertion of part (i) is not true in general even for the open set X. It suffices to consider the single-dimensional map $G(\alpha) = \alpha^3$, which is clearly a P-mapping, but $G'(\alpha) = 3\alpha^2$, i.e. the Jacobian is zero at the zero point.

Combining Propositions 7.9 and 9.2, we obtain similar properties of differentiable M type mappings.

Corollary 9.1. *Let $G : X \to \mathbb{R}^n$ be a continuously differentiable mapping on a rectangle set $X \subseteq \mathbb{R}^n$. Then the following assertions hold:*

(i) If $\nabla G(x)$ is an M-matrix for all $x \in X$, then G is an M-mapping.

(ii) If $\nabla G(x)$ is an M_0-matrix for all $x \in X$, then G is an M_0-mapping.

(iii) If X is open and G is an M_0-mapping, then $\nabla G(x)$ is an M_0-matrix for all $x \in X$.

We now consider the class of strict P-matrices. The next assertion can be viewed as an extension of Proposition 7.5 from affine maps.

Proposition 9.3. *Let X be a rectangle set in $\mathbb{R}^n$ of form (7.4). If $G : X \to \mathbb{R}^n$ is a P_0-mapping, then for any $\varepsilon > 0$, $G+\varepsilon I$ is a strict P-mapping.*

Proof. First we show that $G^{(\varepsilon)} = G + \varepsilon I$ is a P-mapping for each $\varepsilon > 0$. Choose $x', x'' \in X$, $x' \neq x''$, set $I = \{i \mid x'_i \neq x''_i\}$ and fix $\varepsilon > 0$. Since G is a P_0-mapping, there exists an index $k \in I$ such that

$$(x'_k - x''_k)\,[G_k(x') - G_k(x'')] = \max_{1 \leq i \leq n} (x'_i - x''_i)\,[G_i(x') - G_i(x'')] \geq 0,$$

but $x'_k \neq x''_k$ and

$$\varepsilon(x'_k - x''_k)(x'_k - x''_k) > 0.$$

Adding these inequalities gives

$$(x'_k - x''_k)\left[G_k^{(\varepsilon)}(x') - G_k^{(\varepsilon)}(x'')\right] > 0,$$

i.e. $G^{(\varepsilon)}$ is a P-mapping. Choose arbitrary values $0 < \varepsilon'' < \varepsilon'$, then $G^{(\varepsilon'')} = G^{(\varepsilon')} - (\varepsilon' - \varepsilon'')I = G + \varepsilon'' I$ is a P-mapping. It means that $G^{(\varepsilon)}$ is a strict P-mapping. □

Thus, we can obtain a strict P-mapping from an arbitrary P_0-mapping by using a small perturbation. However, in the differentiable case, strict P-mappings ensure the converse of part (i) in Proposition 9.2.

Proposition 9.4. *Let $G : X \to \mathbb{R}^n$ be a continuously differentiable mapping on an open rectangle set $X \subseteq \mathbb{R}^n$. If G is a strict P-mapping, then $\nabla G(x)$ is a P-matrix for all $x \in X$.*

Proof. Since G is a strict P-mapping, there exists $\varepsilon > 0$ such that $H = G - \varepsilon I$ is a P-mapping, hence its Jacobian ∇H is a P_0-matrix. Since $G = H + \varepsilon I$, using Proposition 7.5 gives that $\nabla G(x)$ is a P-matrix. □

Let us consider the standard CP (7.3). For the convenience of the reader, we recall that the problem is to find a point $x^* \in \mathbb{R}^n$ such that

$$x_i^* \geq 0, G_i(x^*) \geq 0, x_i^* G_i(x^*) = 0 \text{ for } i = 1, \dots, n; \tag{9.2}$$

where $G : \mathbb{R}^n_+ \to \mathbb{R}^n$ is a given mapping.

Proposition 9.5. *Suppose that G is a P-mapping. Then CP (9.2) has at most one solution.*

Proof. On the contrary, suppose that there exist at least two distinct solutions x' and x'' of CP (9.2). Then we have

$$G_i(x')(x''_i - x'_i) \geq 0$$

and

$$G_i(x'')(x'_i - x''_i) \geq 0$$

for all $i = 1, \ldots, n$. Addition of these inequalities gives

$$[G_i(x') - G_i(x'')] (x'_i - x''_i) \leq 0 \quad \text{for} \quad i = 1, \ldots, n;$$

i.e. G is not a P-mapping. □

This uniqueness result yields additional properties of CP's with M cost mappings. For example, if all the assumptions of Theorem 7.2 hold, and in addition G is a P-mapping (i.e., it is a continuous M-mapping), then CP (7.3) (or (9.2)) has a unique solution, which coincides with $\min D$ and the Jacobi algorithm from Section 7.3 then converges to $\min D$.

Exercise 9.2. Prove that the assertion of Proposition 9.5 remains true, if we replace CP (9.2) by MCP (7.2) (7.4).

We are now ready to establish the basic existence and uniqueness result.

Theorem 9.1. *Let $G : \mathbb{R}^n_+ \to \mathbb{R}^n$ be a continuous strict P-mapping. Then CP (9.2) has a unique solution.*

Proof. Due to Proposition 9.5, it suffices to show that CP (9.2) is solvable. Given a number $\rho > 0$, set

$$X_\rho = \{x \in \mathbb{R}^n \mid 0 \leq x_i \leq \rho \quad i = 1, \ldots, n\}$$

and consider the following MCP: Find $x^\rho \in X_\rho$ such that

$$(x - x^\rho)^T G(x^\rho) \geq 0 \quad \forall x \in X_\rho.$$

Taking into account Corollary 7.1 and Exercise 9.2, we conclude that this MCP always has the unique solution x^ρ for each $\rho > 0$. We now proceed to show that $\|x^\rho\| < \rho$ for $\rho > 0$ large enough. Assume for contradiction that $\|x^\rho\| \to \infty$ as $\rho \to +\infty$. Choose an arbitrary sequence $\{\rho_s\} \to +\infty$ and set $y^s = x^{\rho_s}$. Also, define the index set $J = \{i \mid y^s_i \to +\infty \quad \text{as} \quad s \to +\infty\}$ and set

$$z^s_i = \begin{cases} y^s_i & \text{if} \quad i \notin J, \\ 0 & \text{if} \quad i \in J; \end{cases}$$

for $i = 1, \dots, n$. Then we have

$$(y_{i_s}^s - z_{i_s}^s)\left[G_{i_s}(y^s) - G_{i_s}(z^s) - \varepsilon(y_{i_s}^s - z_{i_s}^s)\right]$$
$$= \max_{1 \le i \le n} (y_i^s - z_i^s)\left[G_i(y^s) - G_i(z^s) - \varepsilon(y_i^s - z_i^s)\right] > 0$$

for some $\varepsilon > 0$. Since the index set $\{1, \dots, n\}$ is fixed, without loss of generality we can assume that so is i_s, i.e. $i_s = l$. It follows that $l \in J$ and

$$y_l^s G_l(y^s) > \varepsilon(y_l^s)^2 + y_l^s G_l(z^s).$$

Since $\{z^s\}$ is bounded, we have

$$\varepsilon(y_l^s)^2 + y_l^s G_l(z^s) \to +\infty \quad \text{as} \quad s \to \infty,$$

i.e. $y_l^s G_l(y^s) > 0$ for k large enough. But $0 \in X_\rho$ and, in view of Proposition 7.2, this inequality contradicts the definition of y^s. It follows that there exists a number s' such that $\|y^s\| < \rho_s$, hence $0 \le y_i^s < \rho_s$ for $i = 1, \dots, n$, if $s \ge s'$. Take an arbitrary point $x \in \mathbb{R}_+^n$. Then there exists a number $\delta > 0$ such that $0 \le y_i^s + \delta(x_i - y_i^s) < \rho_s$. Applying Proposition 7.2, we see that

$$\left[y_i^s + \delta(x_i - y_i^s) - y_i^s\right] G_i(y^s) \ge 0,$$

or equivalently,

$$(x_i - y_i^s) G_i(y^s) \ge 0$$

for $i = 1, \dots, n$, if $s \ge s'$. Since $y^s \in \mathbb{R}_+^n$, using Proposition 7.1 and Exercise 7.1, we conclude that y^s now solves CP (9.2). □

This existence result allows us to replace the condition $D \neq \emptyset$ in Theorem 7.3 with the strict P property of G. Then, the Jacobi algorithm will converge to the unique solution $\min D$ of CP (7.3). This combination of properties seems very natural for economic applications.

We give additionally existence and uniqueness results, which are similar to those in Section 7.3, under usual (norm) monotonicity assumptions on G, although such properties are mostly used for general VI's; see Chapter 11.

Definition 9.2. Let X be a convex set in $\mathbb{R}^n$. A mapping $G : X \to \mathbb{R}^n$ is said to be

(a) *monotone*, if, for each pair of points $x', x'' \in X$, it holds that

$$(x' - x'')^T \left[G(x') - G(x'')\right] \ge 0;$$

(b) *strictly monotone*, if, for each pair of points $x', x'' \in X$, $x' \neq x''$, it holds that

$$(x' - x'')^T \left[G(x') - G(x'')\right] > 0$$

From the definitions it follows that monotonicity (strict monotonicity) is stronger than the $P_0(P)$ property.

Exercise 9.3. Let X be a set of form (7.4). Prove that $G : X \to \mathbb{R}^n$ is a P_0-(respectively, P-) mapping, if it is monotone (respectively, strictly monotone), but the reverse assertions are not true.

Next, let us consider the affine case when G is defined in (9.1). Then monotonicity becomes closely related with positive (semi)definiteness of A.

Exercise 9.4. Suppose that G is of form (9.1). Prove that G is monotone (respectively, strictly monotone) if and only if A is positive semidefinite (respectively, positive definite).

These assertions are similar to those in Proposition 9.1, hence Exercise 9.3 can be viewed as an extension of Corollary 7.2. It means that CP (9.2) with G being strictly monotone has also at most one solution.

Theorem 9.2. *(i) If $G : X \to E$ is a continuous and monotone mapping and there exists a point $y \geq 0$ such that $G(y) > 0$, then CP (9.2) has a solution.*

(ii) If $G : X \to E$ is a continuous and strictly monotone mapping and there exists a point $y \geq 0$ such that $G(y) \geq 0$, then CP (9.2) has a unique solution.

Proof. Take an arbitrary $x \in \mathbb{R}^n_+$. Then, by monotonicity,

$$(x-y)^T G(x) \geq x^T G(y) - y^T G(y).$$

Since $G(y) > 0$, there exists a number $\beta > 0$ such that the right-hand side of the above inequality is positive, if $x \in \mathbb{R}^n_+ \backslash Y$, where

$$Y = \left\{ x \geq 0 \;\middle|\; \sum_{i=1}^{n} x_i \leq \beta \right\}.$$

Assertion (i) follows now from Theorem 7.1 with $X = \mathbb{R}^n_+$.

In case (ii), it suffices to prove that the conditions of part (i) hold. Fix a number $\beta > 0$. Then Theorem 7.1 with $X = K$ guarantees the existence of a solution $\tilde{x} \in K$ for the following VI:

$$(z - \tilde{x})^T [G(\tilde{x}) - G(y)] \geq 0 \quad \forall z \in K,$$

where

$$K = \left\{ z \geq y \;\middle|\; \sum_{i=1}^{n} z_i = \sum_{i=1}^{n} y_i + \beta \right\}$$

is clearly nonempty, convex, and compact. It follows that

$$z^T[G(\tilde{x}) - G(y)] \geq \tilde{x}^T[G(\tilde{x}) - G(y)] \quad \forall z \in K.$$

Since $\tilde{x} \neq y$, we have

$$(\tilde{x} - y)^T[G(\tilde{x}) - G(y)] > 0,$$

therefore,

$$(z - y)^T[G(\tilde{x}) - G(y)] \geq (\tilde{x} - y)^T[G(\tilde{x}) - G(y)] > 0 \quad \forall z \in K.$$

It follows that $G(\tilde{x}) > 0$, i.e. the conditions of part (i) hold for $y = \tilde{x}$ and (ii) is also true. □

Exercise 9.5. Prove the extensions of Theorems 9.1 and 9.2 for MCP (7.2), (7.4).

9.2 Solution methods for CP's with P properties

There exist several different approaches to compute a solution of CP (9.2) with G being a P type mapping; see e.g. *Facchinei and Pang* (2003). Most of them consist in introduction of an artificial function (or a mapping) which enables one to convert the initial CP into a system of nonlinear equations or into an optimization problem since a great number of effective solution methods were developed for just these classes of problems. In such a way, one can choose a suitable algorithm for solving the initial problem. We consider one of the simplest approaches, which is related with a so-called regularized merit function.

Let us consider CP (9.2) where $G : \mathbb{R}^n_+ \to \mathbb{R}^n$ is a continuously differentiable mapping. Recall that it means that, for each $x \in \mathbb{R}^n_+$, there exists the Jacobian of G, denoted by $\nabla G(x)$, whose entries are partial derivatives $\frac{\partial G_i(x)}{\partial x_j}$ for $i, j = 1, \ldots, n$, and that ∇G is continuous. Next, let us consider the regularized merit function, suggested by *M. Fukushima*, which is defined as follows:

$$\varphi_\alpha(x) = \max_{y \in \mathbb{R}^n_+} \left\{ (x - y)^T G(x) - (2\alpha)^{-1} \|x - y\|^2 \right\},$$

where $\alpha > 0$ is a fixed number. This function can be equivalently re-defined as follows:

$$\begin{aligned} \varphi_\alpha(x) &= \sum_{i=1}^{n} \max_{y_i \geq 0} \left\{ (x_i - y_i) G_i(x) - (2\alpha)^{-1} (x_i - y_i)^2 \right\} \qquad (9.3) \\ &= \sum_{i=1}^{n} \left\{ (x_i - y_i^\alpha) G_i(x) - (2\alpha)^{-1} (x_i - y_i^\alpha)^2 \right\}, \end{aligned}$$

i.e., y_i^α is the unique solution of the i-th one-dimensional quadratic programming problem in (9.3) and it can be computed explicitly: $y_i^\alpha = \max\{0, x_i - \alpha G_i(x)\}$.

Set $y^\alpha(x) = (y_1^\alpha, \dots, y_n^\alpha)^T$, then we have

$$y^\alpha(x) = \pi_+ \left[x - \alpha G(x)\right], \tag{9.4}$$

where $\pi_+(\cdot)$ denotes the projection mapping onto $\mathbb{R}_+^n$; see Section 11.2.

Exercise 9.6. By using the projection properties, prove the formulas for computation of y_i^α and $y^\alpha(x)$.

The basic properties of the function φ_α are collected in the next proposition.

Proposition 9.6. *(i) $\varphi_\alpha(x) \geq 0$ for all $x \geq 0$.*
(ii) The following statements are equivalent:
(a) $\varphi_\alpha(x^*) = 0$ and $x^* \geq 0$,
(b) x^* solves CP (9.2),
(c) $x^* = y^\alpha(x^*)$.

Proof. By definition, for each $x \geq 0$, we have

$$\varphi_\alpha(x) \geq (x - x)^T G(x) - (2\alpha)^{-1}\|x - x\|^2 = 0,$$

hence assertion (i) is true. In case (ii), if x^* solves CP (9.2), then, by Propositions 7.1 and 7.2,

$$(y - x^*)^T G(x^*) \geq 0 \quad \forall y \geq 0, \tag{9.5}$$

hence,

$$\varphi_\alpha(x^*) = \max_{y \geq 0} \left\{(x^* - y)^T G(x^*) - (2\alpha)^{-1}\|x^* - y\|^2\right\} \leq 0,$$

it follows that $\varphi_\alpha(x^*) = 0$ due to (i). Next, in view of (9.4), we can apply Proposition 11.12 (ii) with $x = x^* - \alpha G(x^*)$ and $Y = \mathbb{R}_+^n$ and obtain

$$(y - y^\alpha(x^*))^T \left[y^\alpha(x^*) - x^* + \alpha G(x^*)\right] \geq 0 \quad \forall y \geq 0, \tag{9.6}$$

hence

$$\begin{aligned} \varphi_\alpha(x^*) &= (x^* - y^\alpha(x^*))^T G(x^*) - (2\alpha)^{-1}\|x^* - y^\alpha(x^*)\|^2 \\ &\geq (2\alpha)^{-1}\|x^* - y^\alpha(x^*)\|^2 \end{aligned}$$

and $\varphi_\alpha(x^*) = 0$ now implies $x^* = y^\alpha(x^*)$. However, if $x^* = y^\alpha(x^*)$, then (9.6) implies (9.5), and, on account of Proposition 7.1, x^* solves CP (9.2). Thus, part (ii) is also true. □

The results of Proposition 9.6 show that φ_α is a merit (or gap) function for CP (9.2), i.e. it is equivalent to the optimization problem:

$$\min_{x\geq 0} \to \varphi_\alpha(x). \tag{9.7}$$

However, φ_α is non-convex in general, i.e. problem (9.7) may have local minima which are not solutions of (9.2). Therefore, we should introduce additional conditions ensuring the property that each local minimizer in (9.7) is a solution of CP (9.2). First we note that φ_α is a differentiable function, and, using the formula for derivatives of max type functions (e.g. see *Facchinei and Pang* (2003), Theorem 10.2.1), we have

$$\nabla\varphi_\alpha(x) = G(x) + \left[\nabla G(x) - \alpha^{-1}I\right]^T (x - y^\alpha(x)).$$

Moreover, since the projection mapping is continuous (see e.g. Proposition 11.12 (iii)), (9.4) implies that the mapping $x \mapsto y^\alpha(x)$ is also continuous. Therefore, φ_α is a continuously differentiable function. Let us consider the problem of finding a stationary point $x^* \geq 0$ for (9.7) in the form of VI:

$$(y - x^*)^T \nabla\varphi_\alpha(x^*) \geq 0 \quad \forall y \geq 0. \tag{9.8}$$

Due to Proposition 7.1, it can be equivalently rewritten as CP:

$$x^* \geq 0, \quad \nabla\varphi_\alpha(x^*) \geq 0, \quad (x^*)^T\nabla\varphi_\alpha(x^*) = 0.$$

Theorem 9.3. *Suppose that $\nabla G(x)$ is a P-matrix for all $x \geq 0$. Then problems (9.2), (9.7), and (9.8) have the same solution set.*

Proof. The equivalence between (9.2) and (9.7) is given in Proposition 9.6. Theorem 11.1 yields the implication (9.7) $\Longrightarrow$ (9.8). Suppose that x^* solves (9.8), then

$$(y_i^\alpha - x_i^*)G_i(x^*) + \alpha^{-1}(y_i^\alpha - x_i^*)^2 \geq (y_i^\alpha - x_i^*)\left[\nabla G(x^*)^T(y^\alpha - x^*)\right]_i$$

for $i = 1, \dots, n$, where $y^\alpha = y^\alpha(x^*)$. Applying now (9.6) with $y = x^*$ gives

$$(y_i^\alpha - x^*)\left[\nabla G(x^*)^T(y^\alpha - x^*)\right]_i \leq 0 \quad \text{for} \quad i = 1, \dots, n,$$

but $\nabla G(x^*)$ is a P-matrix and Proposition 7.3 yields $y^\alpha(x^*) = x^*$. On account of Proposition 9.6, it follows that x^* solves (9.7). □

Thus, Theorem 9.3 allows us to replace the initial CP (9.2) with a VI (or CP), which is an optimality condition for the optimization problem (9.7), i.e. its cost mapping is the gradient of the merit function φ_α. Being based on this property, we can suggest the simplest gradient projection algorithm for CP (9.2).

Algorithm (gradient projection). Choose a point $x^0 \in \mathbb{R}^n_+$, numbers $\alpha > 0, \lambda > 0, \beta \in (0,1), \gamma \in (0,1)$, and set $k = 0$.

At the k-th iteration, we have a point $x^k \in \mathbb{R}^n_+$, compute $y^\alpha(x^k)$,

$$z^k = \pi_+ \left[x^k - \lambda \nabla \varphi_\alpha(x^k)\right],$$

and set $d^k = z^k - x^k$. Afterwards, find m as the smallest non-negative integer such that

$$\varphi_\alpha(x^k + \gamma^m d^k) \leq \varphi_\alpha(x^k) - \beta\gamma^m (d^k)^T \nabla \varphi_\alpha(x^k),$$

set $\mu_k = \gamma^m, x^{k+1} = x^k + \mu_k d^k$, and $k = k + 1$.

The convergence result for this method follows from Theorems 9.3 and 13.4.

Theorem 9.4. *Suppose that $\nabla G(x)$ is a P-matrix for all $x \in \mathbb{R}^n_+$, and that the set $L_0 = \left\{x \in \mathbb{R}^n_+ \mid \varphi_\alpha(x) \leq \varphi_\alpha(x^0)\right\}$ is bounded. Then the sequence $\{x^k\}$ generated by the above algorithm is well-defined and converges to a unique solution of CP (9.2).*

Note that the above merit function and algorithm can be also adjusted for MCP (7.2), (7.4).

Chapter 10

Applications

In this chapter, we give examples of applications of the results of the previous chapter concerning theory and methods of CP's with P type mappings.

10.1 Walrasian price equilibrium models

We recall that the general Walrasian price equilibrium model describes an economy with perfect competition. This model was considered in Sections 5.2, 8.2, and 8.3. The economy deals in n commodities, and given a price vector $p \geq 0$, the demand and supply are determined as vectors $D(p)$ and $S(p)$, respectively, hence one can define the excess demand

$$E(p) = D(p) - S(p).$$

The equilibrium price vector is defined to satisfy the complementarity conditions (see (5.3)):

$$p^* \geq 0, E(p^*) \leq 0, (p^*)^T E(p^*) = 0, \tag{10.1}$$

i.e. it solves CP (7.3) with $G(p) = -E(p)$. We denote by P^* the set of equilibrium prices. In Section 8.2, this problem was considered under gross substitutability of (excess) demand, which yielded CP's with Z-mappings. We now consider some additional properties which ensure for $-E$ to be a P_0 (or even an M_0) type mapping.

Definition 10.1. Let X be a box constrained set in $\mathbb{R}^n_+$. A mapping $A : X \to \mathbb{R}^n$ is said to be *positively homogeneous* with degree m, if for each $x \in X$ and for each $\lambda > 0$ such that $\lambda x \in X$, it holds that $A(\lambda x) = \lambda^m A(x)$.

Namely, combining these properties, we obtain that $-E$ is a P_0-mapping.

Proposition 10.1. *Let X be a cone contained in $\mathbb{R}^n_+$. If $A : X \to \mathbb{R}^n$ is positively homogeneous with degree 0 and a GS-mapping, then $-A$ is a P_0-mapping. If in addition A is continuous, then $-A$ is a M_0-mapping.*

Proof. Choose arbitrary points $x', x'' \in X$, $x' \neq x''$. First we consider the case when there exists i such that $x'_i > x''_i$. Set $\theta = \max_{i=1,\dots,n} (x'_i/x''_i) > 1$, then $\theta x'' \geq x'$ and there exists at least one index k such that $x'_k = \theta x''_k$ and, by the GS property,

$$A_k(x') \leq A_k(\theta x'') = A_k(x'').$$

It follows that $x'_k > x''_k$ and

$$-\left[A_k(x') - A_k(x'')\right](x'_k - x''_k) \geq 0.$$

The case when there exists i such that $x'_i < x''_i$ can be considered similarly. It means that $-A$ is a P_0-mapping. If A is continuous, the result follows from Proposition 8.2 and Definition 9.1 (d). □

Observe that positive homogeneity of degree 0 is a very popular property of (excess) demand in classical economic equilibrium models; see e.g. *Morishima* (1964), *Nikaido* (1968), and *Arrow and Hahn* (1971). It follows usually from the insatiability of consumers and also means that money is neutral in such models. Note that the utility functions of consumers from the pure trade market model, given in Section 8.3, satisfy this property and the excess demand functions in (8.11) are positively homogeneous of degree 0, i.e. $-E$ is a M_0-mapping. However, we can not apply directly the existence and uniqueness results from Section 9.1 to these models since the (excess) demand is then undefined for prices with zero coordinates, which also follows from insatiability of consumers. Besides, we recall that, due to the *Euler* theorem, see e.g. *Nikaido* (1968), Lemma 18.4, it holds that

$$\nabla E(p)p = 0 \quad \forall p \in \mathbb{R}^n_>$$

if $E : \mathbb{R}^n_> \to \mathbb{R}^n$ is differentiable and positively homogeneous of degree 0. This means that the Jacobian $\nabla E(p)$ is always degenerate, i.e. $-E$ can not be a P-(or M-) mapping and that the equilibrium price in (10.1) can not be unique. This fact can be deduced directly from the positive homogeneity of degree m, since each solution p^* of (10.1) must be in $\mathbb{R}^n_>$ that implies $E(p^*) = 0$, hence any vector $p = \lambda p^*$ with $\lambda > 0$ is also a solution of (10.1). There are a great number of works devoted to existence results for Walrasian price models; see e.g. *Nikaido* (1968), *Arrow and Hahn* (1971), and *Border* (1985). Some results under positive homogeneity will be given in Section 12.3. For the convenience of the reader, we now give one of such formulations (see Theorem 12.2) here as well.

Proposition 10.2. *Suppose that* $E : \mathbb{R}^n_> \to \mathbb{R}^n$ *is continuous, positively homogeneous with degree* 0*, and satisfies the Walras law, i.e.*

$$p^T E(p) = 0 \quad \forall p \in \mathbb{R}^n_>.$$

Suppose also that for every sequence $\{p^k\} \subset \mathbb{R}^n_>$ *converging to* p*, it holds that*

$$E_i(p^k) = \begin{cases} \to +\infty & \text{if } \; p_i = 0, \\ \geq C > -\infty & \text{if } \; p_i > 0; \end{cases}$$

as $k \to \infty$*. Then problem (10.1) has a solution which also solves the problem*

$$p^* > 0, \quad E(p^*) = 0. \tag{10.2}$$

At the same time, if there are price rigidities in the economy, model (10.1) should be replaced with MCP (8.5), where the feasible box - constrained set X is defined in (8.9) and $G = -E$. For instance, if $0 < \alpha_i < \beta_i < +\infty$ for $i = 1, \dots, n$, then G is finite on X and is usually continuous. The existence result for such a problem follows from Corollary 7.1. Next, following the regularization approach, we can replace the excess demand mapping with the perturbed mapping

$$\tilde{E}(p) = E(p) - \varepsilon p \quad \text{with} \quad \varepsilon > 0, \tag{10.3}$$

then $-\tilde{E}$ becomes a strict P-mapping if E is positively homogeneous with degree 0 and gross substitutable because of Propositions 10.1 and 9.3. Hence, if $\alpha_i > 0$ for $i = 1, \dots, n$, then, due to Theorem 9.1, the perturbed price equilibrium problem will have a unique solution. For ε small enough, this solution is close to the equilibrium price in the initial model.

The results of Section 9.2 can also be applied to Walrasian equilibrium models. Theorem 9.4 says that the gradient projection algorithm, being applied to problem (10.1) requires certain additional assumptions for its convergence. In order to obtain the desired P property, we can again apply the regularization approach and replace the initial price equilibrium problem by its approximation via applying the perturbed mapping $\tilde{E}$ from (10.3). Also, the same algorithm can be applied for finding equilibrium prices under price rigidities. The economic interpretation of the gradient projection algorithm of Section 9.2 is not so easy since it is based upon the artificial merit function, hence it represents quite a complicated control procedure. At the same time, the simplified process, which is based upon the somewhat different direction finding procedure

$$p^{k+1} = \pi_+ \left[p^k + \lambda_k E(p^k)\right], \quad \lambda_k > 0, \tag{10.4}$$

which is also known as the *Walras tâtonnement* process, admits a very clear interpretation. In fact, for each commodity i, (10.4) means that the positive excess demand $E_i(p^k)$ forces the current price p_i to increase and the negative excess demand forces the price to decrease. In both the cases, the absolute value $|E_i(p^{k+1})|$ will decrease and even tends to zero under a suitable choice of the stepsizes. For example, we can choose λ_k as follows:

$$\lambda_k = \frac{\alpha_k}{\|E(p^k)\|}, \sum_{k=0}^{\infty} \alpha_k = \infty, \sum_{k=0}^{\infty} \alpha_k^2 < \infty. \tag{10.5}$$

Then, under the above conditions, algorithm (10.4)–(10.5) generates a sequence of prices $\{p^k\}$ which converges to an equilibrium price. It is known that this convergence result is due to the *revealed preference condition* (10.6) (see *Arrow and Hahn* (1971) and *Nikaido* (1968)), which means geometrically that in the presence of the Walras law, the angle between the excess demand vector $E(p)$ at an arbitrary non-equilibrium price vector p and the direction $p^* - p$ to each equilibrium price is acute, and can be deduced e.g. from Theorem 13.3. Conditions (10.5) determine a compromise between too slow and too rapid convergence of stepsizes to zero and are fulfilled if we set

$$\alpha_k = \frac{\alpha}{k+1} \quad \text{with} \quad \alpha > 0.$$

We now show that the desired condition ensuring the convergence of the tâtonnement process to an equilibrium price vector is fulfilled under certain positive homogeneity and gross substitutability properties. For simplicity, we utilize a somewhat strengthened property (cf. Definition 8.2).

Definition 10.2. Let X be a box constrained set in $\mathbb{R}^n_+$. A mapping $A : X \to \mathbb{R}^n$ is said to be *strict gross substitutable* (or a *strict GS-mapping*) if, for each pair of points $x', x'' \in X$ such that $x' \geq x'', x' \neq x''$, and $I(x', x'') = \{i \mid x'_i = x''_i\}$ is nonempty, there exists an index $k \in I(x', x'')$ with $A_k(x') > A_k(x'')$.

Theorem 10.1. *Suppose that all the assumptions of Proposition 10.2 are fulfilled, and that $E : \mathbb{R}^n_> \to \mathbb{R}^n$ is a strict GS-mapping. Then:*

(i) the equilibrium price vector p^* exists and is defined uniquely up to positive scalar multiples, i.e.

$$P^* = \{p \in \mathbb{R}^n_> \mid p = \lambda p^*, \lambda > 0\};$$

(ii) it holds that

$$(p^*)^T E(p) > 0, \forall p \in \mathbb{R}^n_> \backslash P^*, \forall p^* \in P^*. \tag{10.6}$$

Proof. From Proposition 10.2 it follows that $P^* \neq \emptyset$. Suppose that $p^*, \tilde{p} \in P^*$ and $\tilde{p} \neq \alpha p^*$ for any $\alpha > 0$. Then $E(p^*) = E(\tilde{p}) = 0$ and there exists a number $\beta > 0$ such that $\tilde{p} \geq \beta p^*$ and $\tilde{p}_k = \beta p_k^*$ for at least one k. Since $\tilde{p} \neq \beta p^*$, there exists an index j such that $0 = E_j(\beta p^*) < E_j(\tilde{p}) = 0$, which is a contradiction. Hence, part (i) is true. Next, take $p^* \in P^*$ such that

$$\sum_{i=1}^{n} p_i^* = 1$$

and set

$$\varphi(p) = (p^*)^T E(p).$$

From the properties of E we deduce that the function φ is continuous and bounded below on $\mathbb{R}^n_>$, moreover, it attains the minimal value at a point $\tilde{p}$ of $\mathbb{R}^n_>$. Suppose that $\tilde{p} \notin P^*$, then there exists the maximal number $\alpha > 0$ such that $q = \alpha p^* \leq \tilde{p}$, hence $I(\tilde{p}, q) = \{j \mid \tilde{p}_j = q_j\} \neq \emptyset$, but $q \neq \tilde{p}$. Since E is a strict GS-mapping, there exists an index $j \in I(\tilde{p}, q)$ such that

$$E_j(\tilde{p}) > E_j(q) = E_j(p^*) = 0.$$

Choose $\delta > 0$ small enough and set

$$\tilde{q}_i = \begin{cases} \tilde{p}_i & \text{if } i \neq j, \\ \tilde{p}_i + \delta & \text{if } i = j. \end{cases}$$

Then $\tilde{q} = (\tilde{q}_1, \dots, \tilde{q}_n) \in \mathbb{R}^n_>$ and $\varphi(\tilde{p}) \leq \varphi(\tilde{q})$. For brevity, set $d_i = E_i(\tilde{p}) - E_i(\tilde{q})$. By using the Walras law, we have

$$\sum_{i=1}^{n} d_i \tilde{p}_i = \sum_{i=1}^{n} E_i(\tilde{p})\tilde{p}_i - \sum_{i=1}^{n} E_i(\tilde{q})\tilde{p}_i = -\sum_{i=1}^{n} E_i(\tilde{q})\tilde{p}_i$$

and

$$0 = \sum_{i=1}^{n} \tilde{q}_i E_i(\tilde{q}) = \delta E_j(\tilde{q}) + \sum_{i=1}^{n} \tilde{p}_i E_i(\tilde{q}),$$

hence

$$\sum_{i=1}^{n} d_i \tilde{p}_i = \delta E_j(\tilde{q}),$$

or equivalently,

$$d_j \tilde{p}_j = \delta E_j(\tilde{q}) - \sum_{i \neq j} d_i \tilde{p}_i. \tag{10.7}$$

At the same time, we have

$$\sum_{i=1}^{n} d_i p_i^* \leq 0,$$

or

$$d_j p_j^* \leq -\sum_{i \neq j} d_i p_i^*.$$

Multiplying this inequality by α and subtracting it from (10.7), we obtain

$$d_j(\tilde{p}_j - \alpha p_j^*) \geq \delta E_j(\tilde{q}) - \sum_{i \neq j} d_i(\tilde{p}_i - \alpha p_i^*).$$

Due to Proposition 8.2, $-E$ is a Z-mapping on $\mathbb{R}^n_>$, therefore $d_i \leq 0$ for every $i \neq j$, but $\tilde{p}_i - \alpha p_i^* \geq 0$ for $i = 1, \ldots, n$ and we can choose $\delta > 0$ small enough so that $E_j(\tilde{q}) > 0$ due to continuity of E. It follows that $d_j(\tilde{p}_j - \alpha p_j^*) > 0$, i.e. $\tilde{p}_j \neq \alpha p_j^*$, which is a contradiction. This means that part (ii) is true. □

It is known that the strict GS property of E in Theorem 10.1 can be replaced by the usual GS property, however, the substantiation is then essentially longer; see e.g. *Nikaido* (1968), Section 18.3.

Now we consider a modification of the classical Walrasian model above. Namely, divide all the economical agents into two parts. The first part consists of classical consumers whose total demand mapping $D : \mathbb{R}^n_> \rightarrow \mathbb{R}^n$ is continuous, gross substitutable, and positive homogeneous with degree 0. They may also possess some endowments of commodities and the total endowments are given by the vector $S \in \mathbb{R}^n_+$. An example of such economic agents was presented in the pure trade market model from Section 8.3. Besides, we suppose that the market involves a number of "single-dimensional" consumers and producers so that each of them either demands or supplies a single commodity, and these values may only depend on the price of just this commodity. Then, without loss of generality we can suppose that the behavior of the second part can be described by the diagonal excess demand mapping

$$B(p) = (B_1(p_1), \ldots, B_n(p_n))^T$$

and that each function $B_i : \mathbb{R}_+ \rightarrow \mathbb{R}$ is non-increasing and continuous for $i = 1, \ldots, n$. Then we can define the total excess demand mapping

$$E(p) = D(p) + B(p) - S, \tag{10.8}$$

and the equilibrium price vector can be defined from (10.1). The market structure described need not be closed, i.e. the Walras law does not hold in general. It follows from Proposition 10.1 that $-D$ is a M_0-mapping, moreover, $-B$ is monotone by definition, i.e. it is a P-mapping. It was noticed in Section 9.1 that the classes of P_0- and P-mappings are not additive, but in this special case we can establish P type properties of the mapping $G = -E$.

Definition 10.3. A function $\varphi : \mathbb{R} \to \mathbb{R}$ is said to be strongly decreasing with modulus $\tau > 0$, if, for each pair of numbers $x', x'' \in \mathbb{R}$, it holds that

$$[\varphi(x') - \varphi(x'')]\,(x' - x'') \leq -\tau(x' - x'')^2.$$

Together with the usual monotonicity properties of the functions B_i, we shall describe the behavior of the second part of economic agents with the help of such strengthened conditions. As a result, we obtain strengthened P properties of the mapping G.

Proposition 10.3. *If each function $B_i : \mathbb{R}_+ \to \mathbb{R}$ is non-increasing (respectively, decreasing, strongly decreasing), then the mapping $-E : \mathbb{R}^n_> \to \mathbb{R}^n$, defined in (10.8), is a P_0-mapping (respectively, a P-mapping, a strict P-mapping).*

Exercise 10.1. Prove the assertion of Proposition 10.3.

Proposition 10.3 enables us to apply the results of Chapter 9 to the above equilibrium model. Let us consider its extended form (8.5) with the feasible set X being defined in (8.9) and $G = -E$. If $0 < \alpha_i < \beta_i < +\infty$ for $i = 1, \ldots, n$, then, on account of Corollary 7.1, we see that the equilibrium price vector exists, and if additionally B_i is decreasing for $i = 1, \ldots, n$, then it is defined uniquely because of Proposition 9.5. Next, applying Theorem 9.1, we see that MCP (8.5), (8.9) has a unique solution if $\alpha_i > 0$ and B_i is strongly decreasing for each $i = 1, \ldots, n$. Moreover, the algorithm from Section 9.2 can be utilized to find equilibrium prices.

10.2 Oligopolistic equilibrium models

We now describe applications of the results of Chapter 9 to the imperfectly competitive equilibrium model from Section 5.4. In this model, there are n firms supplying a homogeneous product. Unlike Section 8.4, we consider here the Cournot oligopoly. If the i-th firm supplies x_i units of the product, then its profit is defined by

$$f_i(x) = x_i p(\sigma_x) - h_i(x_i),$$

where $x = (x_1, \dots, x_n)$, $h_i : \mathbb{R}_+ \to \mathbb{R}$ is the cost function, $p : \mathbb{R}_+ \to \mathbb{R}$ is the inverse demand (or price) function of the market, and $\sigma_x = \sum_{i=1}^{n} x_i$ is the total supply in the market. As indicated in Section 5.4, the solution x^* can be defined as the Nash equilibrium point of the n-person non-cooperative game, where the i-th player has the utility function f_i and the strategy set $\mathbb{R}_+$; see (5.19).

We shall consider this problem under the following blanket assumptions. Suppose that the cost functions h_i are convex, the inverse demand function p is non-increasing, and that the industry revenue function $\mu(\sigma) = \sigma p(\sigma)$ is concave on $\mathbb{R}_+$. These assumptions conform to the usually accepted economic behavior; see e.g. *Okuguchi and Szidarovszky* (1990). Besides, for the sake of simplicity, we also suppose that all the functions p and h_i are twice continuously differentiable.

The above condition on μ was proposed by *F.H. Murphy, H.D. Sherali, and A.L. Soyster*, they noticed that it implies the concavity of each utility function f_i in x_i.

Lemma 10.1. *The i-th firm profit function f_i is concave in x_i for $i = 1, \dots, n$.*

Proof. Set $\varphi(\alpha) = \alpha p(\alpha + \beta)$, then, by assumption,

$$\begin{aligned} \varphi''(\alpha) &= 2p'(\alpha + \beta) + \alpha p''(\alpha + \beta) \\ &= [(\alpha + \beta)p(\alpha + \beta)]'' - \beta p''(\alpha + \beta), \end{aligned}$$

also, we have $p'(\alpha + \beta) \leq 0$. If we suppose that $\varphi''(\alpha) > 0$, then, for $\beta > 0$, we obtain

$$\alpha p''(\alpha + \beta) > 0 \quad \text{and} \quad -\beta p''(\alpha + \beta) > 0,$$

a contradiction. Since h_i convex, we now conclude that f_i is concave in x_i on $\mathbb{R}_+$. □

This property enables us to replace the Nash equilibrium problem with CP (9.2), where

$$G_i(x) = h_i'(x_i) - p(\sigma_x) - x_i p'(\sigma_x), \quad i = 1, \dots, n, \tag{10.9}$$

or with problem (7.2), where $X = \mathbb{R}^n_+$; see (5.20) and (5.21). The equivalence result follows now from Corollary 11.2.

Theorem 10.2. *Under the above assumptions, the Nash equilibrium problem (5.19) is equivalent to CP (9.2) (or (5.20)), (10.9).*

Exercise 10.2. Prove the assertion of Theorem 10.2.

We intend to establish P type properties of the mapping G. Its Jacobian $\nabla G(x)$ can be written as follows:

$$\nabla G(x) = A_n(x) + H(x),$$

where $H(x)$ is the $n \times n$ diagonal matrix with diagonal entries $h_i''(x_i)$ for $i = 1, \ldots, n$, and

$$A_n(x) = \begin{pmatrix} \beta + \alpha_1 & \alpha_1 & \alpha_1 & \cdots & \alpha_1 \\ \alpha_2 & \beta + \alpha_2 & \alpha_2 & \cdots & \alpha_2 \\ \vdots & \vdots & \vdots & \ddots & \vdots \\ \alpha_n & \alpha_n & \alpha_n & \cdots & \beta + \alpha_n \end{pmatrix},$$

where we set $\beta = -p'(\sigma_x)$ and $\alpha_i = -p'(\sigma_x) - x_i p''(\sigma_x)$ for brevity. It follows that it is sufficient to consider only the principal minors $\det A_k(x)$ of the matrix $A_n(x)$.

Lemma 10.2. *It holds that*

$$\det A_k(x) = (-p'(\sigma_x))^{k-1} \left[-(k+1)p'(\sigma_x) - \left(\sum_{i=1}^{k} x_i \right) p''(\sigma_x) \right].$$

Proof. By definition,

$$\det A_k(x) = \begin{vmatrix} \beta + \alpha_1 & \alpha_1 & \alpha_1 & \cdots & \alpha_1 \\ \alpha_2 & \beta + \alpha_2 & \alpha_2 & \cdots & \alpha_2 \\ \vdots & \vdots & \vdots & \ddots & \vdots \\ \alpha_k & \alpha_k & \alpha_k & \cdots & \beta + \alpha_k \end{vmatrix}.$$

Adding all the rows to the first one and subtracting the first column from others yields

$$\det A_k(x) = \begin{vmatrix} \beta + \sum_{i=1}^{k} \alpha_i & 0 & 0 & \cdots & 0 \\ \alpha_2 & \beta & 0 & \cdots & 0 \\ \vdots & \vdots & \vdots & \ddots & \vdots \\ \alpha_k & 0 & 0 & \cdots & \beta \end{vmatrix} = \beta^{k-1}(\beta + \sum_{i=1}^{k} \alpha_i).$$

Hence,

$$\det A_k(x) = (-p'(\sigma_x))^{k-1} \left[-(k+1)p'(\sigma_x) - \left(\sum_{i=1}^{k} x_i \right) p''(\sigma_x) \right].$$

$\square$

Therefore, non-negativity of the principal minors ensures the desired properties.

Proposition 10.4. *(i)* $G : \mathbb{R}^n_+ \to \mathbb{R}^n$ *in (10.9) is a* P_0*-mapping.*

(ii) If in addition $h_i, i = 1, \dots, n$*, are strictly convex, then* G *is a* P*-mapping.*

(iii) If in addition $h_i, i = 1, \dots, n$*, are strongly convex, then* G *is a strict* P*-mapping.*

Proof. By assumption, $p'(\sigma) \leq 0$. Fix $x \in \mathbb{R}^n_+$. If $p''(\sigma_x) \leq 0$, then it follows from Lemma 10.2 that all the principal minors of $A_n(x)$ are non-negative. Otherwise, we have $p''(\sigma_x) > 0$ and

$$\det A_k(x) = (-p'(\sigma_x))^{k-1}$$
$$\times \left[-(k-1)p'(\sigma_x) - \mu''(\sigma_x) + \left(\sum_{i=k+1}^{n} x_i \right) p''(\sigma_x) \right] \geq 0,$$

because $\mu''(\sigma_x) \leq 0$ by assumption. Again, all the principal minors of $A_n(x)$ are non-negative. Since $A_n(x) = \nabla F(x)$, where

$$F_i(x) = -p(\sigma_x) - x_i p'(\sigma_x), \quad i = 1, \dots, n, \tag{10.10}$$

$F : \mathbb{R}^n_+ \to \mathbb{R}^n$ is a P_0-mapping because of Proposition 9.2. Moreover, all the derivatives $h''_i(x_i)$ are non-negative. Hence, assertion (i) is true. In case (ii) (respectively, (iii)), we see that each h'_i is strictly monotone (respectively, strongly monotone) on account of Proposition 11.3. It follows that these assertions are true. □

By imposing additional conditions on the inverse demand and industry revenue functions we can also obtain strengthened P properties of G.

Proposition 10.5. *(i) If* $p'(\sigma) < 0$ *and either* $\mu''(\sigma) < 0$ *or* $p''(\sigma) \leq 0$ *for all* $\sigma \geq 0$*, then* $\nabla G(x)$ *is a* P*-matrix for every* $x \geq 0$*.*

(ii) If there exists $\delta > 0$ *such that* $p'(\sigma) \leq -\delta$ *and either* $\mu''(\sigma) < -\delta$ *or* $p''(\sigma) \leq 0$ *for all* $\sigma \geq 0$*, then* G *is a strict* P*-mapping.*

Proof. Since the matrix $H(x)$ is diagonal with non-negative entries, it suffices to establish the above properties for the mapping F from (10.10) and for its Jacobian. In case (i), following the proof of Proposition 10.4, we see from Lemma 10.2 that all the principal minors of $\nabla F(x) = A_n(x)$ are positive. By definition, $\nabla F(x)$ is a P-matrix and so is $\nabla G(x)$. In case (ii), fix $\varepsilon \in (0, \delta)$, then

$$\det(A_k(x) - \varepsilon I)$$

$$= \begin{vmatrix} \beta+\alpha_1-\varepsilon & \alpha_1 & \alpha_1 & \dots & \alpha_1 \\ \alpha_2 & \beta+\alpha_2-\varepsilon & \alpha_2 & \dots & \alpha_2 \\ \vdots & \vdots & \vdots & \ddots & \vdots \\ \alpha_k & \alpha_k & \alpha_k & \dots & \beta+\alpha_k-\varepsilon \end{vmatrix}$$

$$= (\beta-\varepsilon)^{k-1}\left(\beta-\varepsilon+\sum_{i=1}^{k}\alpha_i\right).$$

Hence,

$$\det(A_k(x)-\varepsilon I) = (-p'(\sigma_x)-\varepsilon)^{k-1} \times \left[-(k+1)p'(\sigma_x)-\varepsilon-\left(\sum_{i=1}^{k}x_i\right)p''(\sigma_x)\right].$$

If $p''(\sigma_x) \leq 0$, it follows that $\det(A_k(x)-\varepsilon I) > 0$. By definition, F is a strict P-mapping, and so is G. □

These properties enable us to establish several existence and uniqueness results for the oligopolistic equilibrium problem by utilizing the corresponding results from Section 9.1.

Theorem 10.3. *Suppose that at least one of the following conditions holds:*
(a) $h_i, i=1,\dots,m$, are strictly convex;
(b) $p'(\sigma)<0$ and either $\mu''(\sigma)<0$ or $p''(\sigma)\leq 0$ for all $\sigma \geq 0$.
Then CP (9.2), (10.9) has at most one solution.

The result follows from Propositions 10.4 (ii), 10.5 (i), and 9.5.

Theorem 10.4. *Suppose that at least one of the following conditions holds:*
(a) $h_i, i=1,\dots,m$, are strongly convex;
(b) there exists $\delta>0$ such that $p'(\sigma)\leq -\delta$ and either $\mu''(\sigma)\leq -\delta$ or $p''(\sigma)\leq 0$ for all $\sigma\geq 0$.
Then CP (9.2), (10.9) has a unique solution.

The result follows from Propositions 10.4 (iii), 10.5 (ii), and Theorem 9.1.

By introducing an additional coercivity condition, we can somewhat strengthen the above results.

Definition 10.4. Industry output is said to be bounded if there exists a compact subset K of $\mathbb{R}^n_+$ such that

$$G_i(x) > 0 \quad i=1,\dots,n \quad \text{for every} \quad x \in \mathbb{R}^n_+ \backslash K.$$

This condition was suggested by *C.D. Kolstad and L. Mathiesen* and has rather explicit economic sense. In fact, this means that there exists an output level for which marginal profits are negative for all producers.

Theorem 10.5. *Suppose that industry output is bounded. Then CP (9.2), (10.9) is solvable. If in addition the conditions of Theorem 10.3 are fulfilled, then CP (9.2), (10.9) has a unique solution.*

Proof. Without loss of generality we can suppose that $0 \in K$, then condition (7.5) holds, and Theorem 7.1 now implies that problem (7.2), (10.9) with $X = \mathbb{R}^n_+$ is solvable. □

All the above results can be adjusted for the case when outputs of producers are bounded, i.e., $x_i \in [0, \beta_i]$, $\beta_i < +\infty$ for $i = 1, \dots, n$. Then CP (9.2), (10.9) can be clearly replaced with MCP (7.2), where

$$X = [0, \beta_1] \times \cdots \times [0, \beta_n].$$

Exercise 10.3. Extend Theorems 10.2–10.5 for the case of bounded outputs.

These results allow us to apply the gradient projection algorithm, described in Section 9.2, to finding solutions of the oligopolistic equilibrium problems. Although this algorithm relies upon the computation of the gradient of the artificial gap function φ_α (see (9.3)), its descent direction at x^k is close to the vector $-G(x^k)$ where $G_i(x^k)$ is defined in (10.9), and can be also viewed as a dynamical game process, which strengthens stability properties of the usual projection process:

$$x^{k+1} = \pi_+ \left[x^k - \lambda_k G(x^k) \right],$$

i.e., $x_i^{k+1} = \max\{x_i^k - \lambda_k G_i(x^k), 0\}$ for $i = 1, \dots, n$. In fact, given an output vector x^k, the difference between the marginal profit $p(\sigma_{x^k}) + x_i^k p'(\sigma_{x^k})$ and the marginal cost $h_i'(x_i^k)$ is determined. If this value is positive (negative), then the output of the i-th producer increases (decreases) in the simplified process, which corresponds to the natural behavior of players. Under certain additional assumptions, this somewhat modified process will converge to a solution. Behavior of other dynamical game processes are described in the book by *Okuguchi and Szidarovszky* (1990).

Exercise 10.4. Give the sufficient conditions of convergence of the algorithm of Section 9.2 for CP (9.2), (10.9).

Of course, there exist specialized tools for investigation and solution of particular classes of economic equilibrium problems. Nevertheless, the rather general approach presented can be always considered as a good starting point for developing very efficient methods.

Part III

VARIATIONAL INEQUALITIES

The variational inequalities constitute a very general class of problems in Nonlinear Analysis and are very suitable for formulation of various equilibrium models. In this part, we present relationships between variational inequalities and other basic nonlinear problems, existence and uniqueness results, basic solution methods and their applications to the previous equilibrium models.

Chapter 11

Theory of Variational Inequalities

In this chapter, we consider some elements of theory of variational inequality problems with continuous single-valued mappings under a finite-dimensional space setting.

11.1 Variational inequalities and related problems

In this section, we give some facts from the theory of variational inequality problems and their relations with other problems of Nonlinear Analysis.

Let X be a nonempty, closed and convex subset of a finite-dimensional Euclidean space E, $G : X \to E$ a continuous mapping. The usual *variational inequality problem* (VI for short) is the problem of finding a point $x^* \in X$ such that

$$(x - x^*)^T G(x^*) \geq 0 \quad \forall x \in X; \tag{11.1}$$

see Figure 11.1. Its solution set will be denoted by X^*.

We first recall definitions of monotonicity for mappings.

Definition 11.1. Let X be a convex set in E and let $Q : X \to E$ be a mapping. The mapping Q is said to be

(a) *strongly monotone* on X with constant $\tau > 0$ if for each pair of points $x, y \in X$, it holds that

$$(x - y)^T [Q(x) - Q(y)] \geq \tau \|x - y\|^2;$$

(b) *strictly monotone* on X if for all distinct $x, y \in X$, it holds that

$$(x - y)^T [Q(x) - Q(y)] > 0;$$

Figure 11.1:

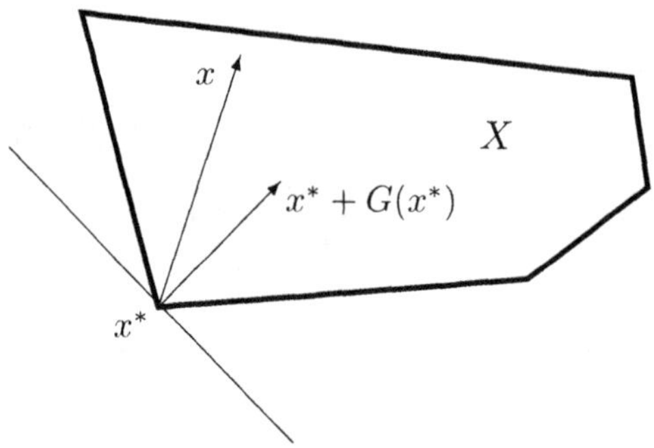

(c) *monotone* on X if for each pair of points $x, y \in X$, it holds that

$$(x - y)^T [Q(x) - Q(y)] \geq 0.$$

It follows from the definitions that the following implications hold:

$$(a) \Longrightarrow (b) \Longrightarrow (c).$$

Observe that all these properties are additive. In addition, we give the known monotonicity criteria for continuously differentiable mappings; see e.g. *Facchinei and Pang* (2003), Proposition 2.3.2.

Proposition 11.1. *Let Y be an open convex subset of X and let $Q : X \to E$ be continuously differentiable on Y.*

(i) Q is monotone on Y if and only if ∇Q is positive semidefinite on Y;

(ii) Q is strictly monotone on Y if ∇Q is positive definite on Y;

(iii) Q is strongly monotone on Y with constant τ if and only if it holds that

$$(p)^T \nabla Q(x) p \geq \tau \|p\|^2 \qquad \forall p \in E, x \in Y.$$

Note that the Jacobian of a differentiable strictly monotone mapping need not be positive definite.

The simplest example of VI is the problem of solving a system of equations. It is easy to see that if $X = E$ in (11.1), then VI (11.1) is equivalent to the problem of finding a point $x^* \in E$ such that

$$G(x^*) = 0.$$

If the mapping G is affine, i.e., $G(x) = Ax - b$, then the above problem is equivalent to the classical system of linear equations

$$Ax^* = b.$$

The case when X is a convex cone in E was considered in Part II. Then VI (11.1) is equivalent to the *complementarity problem* (CP for short):

$$x^* \in X, G(x^*) \in X', (x^*)^T G(x^*) = 0, \tag{11.2}$$

where $X' = \{z \in E \mid x^T z \geq 0 \quad \forall x \in X\}$ is the dual cone to X. This problem can be viewed as a particular case of VI, as showed in Proposition 7.1.

Next, let X be again an arbitrary convex closed set in E and let T be a continuous mapping from X into itself. The *fixed point problem* is to find a point $x^* \in X$ such that

$$x^* = T(x^*). \tag{11.3}$$

This problem can be also converted into a VI format.

Proposition 11.2. *If the mapping G is defined by*

$$G(x) = x - T(x), \tag{11.4}$$

then problem (11.1) coincides with problem (11.3).

Proof. If x^* solves problem (11.3), then clearly $G(x^*) = 0$ and x^* solves problem (11.1), (11.4). Conversely, let x^* solve problem (11.1), (11.4). Then $T(x^*) \in X$ and letting $x = T(x^*)$ in (11.1) gives $-\|x^* - T(x^*)\|^2 \geq 0$, i.e. $x^* = T(x^*)$. □

The reverse transformation based on the projection mapping is described in Proposition 11.13.

Now, we consider the well-known optimization problem. Let $f : X \to \mathbb{R}$ be a real-valued function. Then we can define the following *optimization problem* of finding a point $x^* \in X$ such that

$$f(x^*) \leq f(x) \quad \forall x \in X,$$

or briefly,

$$\min \to \{f(x) \mid x \in X\}. \tag{11.5}$$

We denote by X_f the solution set of this problem.

Recall the definitions of convexity type properties for functions.

Definition 11.2. Let X be a convex set in E and let $\varphi : X \to \mathbb{R}$ be a differentiable function. The function φ is said to be

(a) *strongly convex* on X with constant $\tau > 0$ if for each pair of points $x, y \in X$ and for all $\alpha \in [0, 1]$, it holds that

$$\varphi(\alpha x + (1-\alpha)y) \leq \alpha\varphi(x) + (1-\alpha)\varphi(y) - 0.5\alpha(1-\alpha)\tau\|x-y\|^2;$$

(b) *strictly convex* on X if for all distinct $x, y \in X$ and for all $\alpha \in (0, 1)$, it holds that

$$\varphi(\alpha x + (1-\alpha)y) < \alpha\varphi(x) + (1-\alpha)\varphi(y);$$

(c) *convex* on X if for each pair of points $x, y \in X$ and for all $\alpha \in [0, 1]$, it holds that

$$\varphi(\alpha x + (1-\alpha)y) \leq \alpha\varphi(x) + (1-\alpha)\varphi(y).$$

The function $\varphi : X \to \mathbb{R}$ is said to be *strongly concave* with constant τ (respectively, *strictly concave, concave*) on X, if the function $-\varphi$ is strongly convex with constant τ (respectively, strictly convex, convex) on X.

It follows directly from the definitions that the following implications hold:

$$(a) \Longrightarrow (b) \Longrightarrow (c).$$

The reverse assertions are not true in general.

We now state the relationships between convexity of functions and (generalized) monotonicity of their gradients; see e.g. *Polyak* (1983), Chapter 1, Section 1.4.

Proposition 11.3. *Let Y be an open convex subset of X. A differentiable function $f : X \to \mathbb{R}$ is strongly convex with constant τ (respectively, strictly convex, convex) on Y, if and only if its gradient map $\nabla f : X \to E$ is strongly monotone with constant τ (respectively, strictly monotone, monotone) on Y.*

Similarly, we recall the basic property of the convex differentiable functions; see e.g. *Polyak* (1983), Chapter 1, Section 1.4, Lemma 3.

Proposition 11.4. *Let Y be an open convex subset of X. A differentiable function $f : X \to \mathbb{R}$ is convex on Y, if and only if for each point $x \in Y$, we have*

$$f(y) \geq f(x) + (y-x)^T \nabla f(x) \quad \forall y \in Y.$$

We now give the well-known optimality condition for problem (11.5).

Theorem 11.1. *Suppose that $f : \mathbb{R}^n \to \mathbb{R}$ is a differentiable function. Then:*

(i) $X_f \subseteq X^*$, i.e., each solution of (11.5) is a solution of VI (11.1), where

$$G(x) = \nabla f(x); \tag{11.6}$$

(ii) if f is convex and G is defined by (11.6), then $X^* \subseteq X_f$.

Proof. Part (ii) follows directly from Proposition 11.4. In case (i), assume, for contradiction, that there exists $x^* \in X_f \backslash X^*$, i.e., there is a point $y \in X$ such that

$$(y - x^*)^T \nabla f(x^*) < 0.$$

Then, for $\alpha > 0$ small enough, we must have $y_\alpha = x^* + \alpha(y - x^*) = \alpha y + (1-\alpha)x^* \in X$ and

$$f(y_\alpha) = f(x^*) + \alpha(y - x^*)^T \nabla f(x^*) + o(\alpha) < f(x^*),$$

i.e., $x^* \notin X_f$, which is a contradiction. □

Thus, optimization problem (11.5) can be reduced to VI (11.1) with the monotone underlying mapping G if the function f in (11.5) possesses the corresponding convexity property. However, VI which expresses the optimality condition in optimization enjoys additional properties in comparison with the usual VI. For instance, if f is twice continuously differentiable, then its Hessian matrix $\nabla^2 f = \nabla G$ is symmetric. Conversely, if the mapping $\nabla G : E \to E \times E$ is symmetric, then for any fixed y there exists the function

$$f(x) = \int_0^1 (x-y)^T G(v + \tau(x-y)) d\tau$$

such that (11.6) holds. It is obvious that the Jacobian ∇G of the mapping G in (11.1) is in general asymmetric.

Next, consider the case of the convex optimization problem (11.5). In other words, let the function f be convex and differentiable. Then, according to Theorem 11.1, (11.5) is equivalent to (11.1) with G being defined in (11.6). Due to Proposition 11.3, the mapping ∇f is monotone.

Saddle point problems are closely related to optimization as well as to noncooperative game problems. Let U be a convex closed set in $\mathbb{R}^l$ and let V be a convex closed set in $\mathbb{R}^m$. Suppose that $L : \mathbb{R}^l \times \mathbb{R}^m \to \mathbb{R}$ is a differentiable convex-concave function, i.e., $L(\cdot, v)$ is convex for each $v \in V$ and $L(u, \cdot)$ is concave for each $u \in U$. The *saddle point problem* is to find a pair of points $u^* \in U$, $v^* \in V$ such that

$$L(u^*, v) \leq L(u^*, v^*) \leq L(u, v^*) \quad \forall u \in U, \forall v \in V. \tag{11.7}$$

In particular, the problem of solving a zero-sum two-person game is written in the form (11.7). Set $n = l + m$, $X = U \times V$ and define the mapping $G : \mathbb{R}^n \to \mathbb{R}^n$ as follows:

$$G(x) = G(u, v) = \begin{pmatrix} \nabla_u L(u, v) \\ -\nabla_v L(u, v) \end{pmatrix}. \tag{11.8}$$

From Theorem 11.1 we now obtain the following equivalence result.

Corollary 11.1. *Problems (11.7) and (11.1), (11.8) are equivalent.*

It should be noted that G in (11.8) is also monotone; see Corollary 11.3. Saddle point problems are proved to be a useful tool for "eliminating" functional constraints in optimization. Let us consider the optimization problem

$$\min \to \{f_0(u) \mid u \in D\}, \tag{11.9}$$

where

$$D = \{u \in U \mid f_i(u) \leq 0 \quad i = 1, \dots, m\}, \tag{11.10}$$

$f_i : \mathbb{R}^l \to \mathbb{R}$, $i = 0, \dots, m$ are convex differentiable functions,

$$U = \{u \in \mathbb{R}^l \mid u_j \geq 0 \quad \forall j \in J\}, J \subseteq \{1, \dots, l\}. \tag{11.11}$$

Then we can define the *Lagrange* function associated to problem (11.9) – (11.11) as follows:

$$L(u, v) = f_0(u) + \sum_{i=1}^{m} v_i f_i(u). \tag{11.12}$$

To obtain the relationships between problems (11.9) – (11.11) and (11.7), (11.12), we need certain constraint qualification conditions. Namely, consider the following assumption.

(C) *Either all the functions f_i, $i = 1, \dots, m$ are affine, or there exists a point $\bar{u} \in U$ such that $f_i(\bar{u}) < 0$ for all $i = 1, \dots, m$.*

Proposition 11.5. *(Karush-Kuhn-Tucker; e.g. see Nikaido (1968), Chapter 1, Theorems 3.16 and 3.17)*

(i) If (u^*, v^*) is a saddle point of the function L in (11.12) with $V = \mathbb{R}^m_+$, then u^* is a solution to problem (11.9) – (11.11).

(ii) If u^* is a solution to problem (11.9) – (11.11) and condition (C) holds, then there exists a point $v^* \in V = \mathbb{R}^m_+$ such that (u^*, v^*) is a solution to the saddle point problem (11.7), (11.12).

By using Corollary 11.1, we now see that optimization problem (11.9) – (11.11) can be replaced by VI (11.1) (or equivalently, by CP (11.2) since X is a convex cone), where $X = U \times V$, $V = \mathbb{R}^m_+$, $f(u) = (f_1(u), \dots, f_m(u))^T$, and

$$G(x) = \begin{pmatrix} \nabla f_0(u) + \sum_{i=1}^m v_i \nabla f_i(u) \\ -f(u) \end{pmatrix} \tag{11.13}$$

with G being monotone; see Corollary 11.3.

Similarly, we can convert VI with functional constraints into VI (or CP) with simple constraints. Let us consider the following problem of finding $u^* \in D$ such that

$$(u - u^*)^T F(u^*) \ge 0 \quad \forall u \in D, \tag{11.14}$$

where $F : \mathbb{R}^l \to \mathbb{R}^l$ is a continuous mapping, D is the same as in (11.10), (11.11).

Proposition 11.6. *(i) If $x^* = (u^*, v^*)$ is a solution to (11.1) with*

$$\begin{aligned} &X = U \times V, V = \mathbb{R}^m_+, \\ &G(x) = G(u, v) = \begin{pmatrix} F(u) + \sum_{i=1}^m v_i \nabla f_i(u) \\ -f(u) \end{pmatrix}, \end{aligned} \tag{11.15}$$

then u^* is a solution to problem (11.14), (11.10), (11.11).

(ii) If condition (C) holds and u^* is a solution to problem (11.14), (11.10), (11.11), then there exists a point $v^* \in V = \mathbb{R}^m_+$ such that (u^*, v^*) is a solution to (11.1), (11.15).

Proof. (i) Let (u^*, v^*) be a solution to (11.1), (11.15), or equivalently, to the following system:

$$\begin{aligned} (u - u^*)^T [F(u^*) + \sum_{i=1}^m v_i^* \nabla f_i(u^*)] \ge 0 \quad \forall u \in U, \\ (v - v^*)^T [-f(u^*)] \ge 0 \quad \forall v \in V = \mathbb{R}^m_+. \end{aligned} \tag{11.16}$$

The last relation implies $u^* \in D$ and $[f(u^*)]^T v^* = 0$, so that, using the first relation in (11.16) and Proposition 11.4, we have

$$\begin{aligned} 0 &\leq (u - u^*)^T F(u^*) + \sum_{i=1}^{m} v_i^* (f_i(u) - f_i(u^*)) \\ &\leq (u - u^*)^T F(u^*) \quad \forall u \in D, \end{aligned}$$

i.e., u^* is a solution to problem (11.14), (11.10), (11.11).

(ii) If u^* is a solution to problem (11.14), (11.10), (11.11), it is a solution to the following convex optimization problem

$$\min \to \{u^T F(u^*) \mid u \in D\},$$

and due to Proposition 11.5 (ii), there exists a point $v^* \in V = \mathbb{R}^m_+$ such that (u^*, v^*) is a saddle point of the function $\tilde{L}(u, v) = u^T F(u^*) + \sum_{i=1}^{m} v_i f_i(u)$. Now the left inequality in (11.7) with $L = \tilde{L}$ implies the second inequality in (11.16), whereas the right inequality in (11.7) with $L = \tilde{L}$ implies that u^* is a solution to the following convex optimization problem:

$$\min \to \{u^T F(u^*) + \sum_{i=1}^{m} v_i^* f_i(u) \mid u \in U\},$$

which is equivalent to the first relation in (11.16) due to Theorem 11.1. Therefore, (u^*, v^*) is a solution to (11.1), (11.15). □

We can specialize the above results for various kinds of constraints. For instance, we establish similar results for the case when all the constraint functions are affine. Let us consider the problem of finding $u^* \in \tilde{D} = D \bigcap B$ such that

$$(u - u^*)^T F(u^*) \geq 0 \quad \forall u \in \tilde{D}, \tag{11.17}$$

where $F : \mathbb{R}^l \to \mathbb{R}^l$ is a continuous mapping, D is the same as in (11.10), (11.11),

$$B = \{u \in \mathbb{R}^l \mid h_i(u) = 0 \quad i = 1, \ldots, k\}, \tag{11.18}$$

$h_i : \mathbb{R}^l \to \mathbb{R}$, $i = 1, \ldots, k$ are continuous functions.

Proposition 11.7. *Suppose f_i and h_i are affine, i.e. $f_i(u) = (a^i)^T u - \alpha_i$ for $i = 1, \ldots, m$, and $h_i(u) = (b^i)^T u - \beta_i$ for $i = 1, \ldots, k$.*

(i) If $x^* = (u^*, v^*, w^*)$ *is a solution to (11.1) with*

$$\begin{aligned} X &= U \times \mathbb{R}^m_+ \times \mathbb{R}^k, \\ G(x) &= G(u,v,w) \\ &= \begin{pmatrix} F(u) + \sum_{i=1}^m v_i a^i + \sum_{j=1}^k w_j b^j \\ -f(u) \\ -h(u) \end{pmatrix}, \end{aligned} \tag{11.19}$$

$h(u) = (h_1(u), \ldots, h_k(u))^T$, *then* u^* *is a solution to problem (11.17), (11.18), (11.10), (11.11).*

(ii) If u^* *is a solution to problem (11.17), (11.18), (11.10), (11.11), then there exist points* $v^* \in \mathbb{R}^m_+$ *and* $w^* \in \mathbb{R}^k$ *such that* (u^*, v^*, w^*) *is a solution to (11.1), (11.19).*

This result can be deduced rather easily from Proposition 11.6.

Exercise 11.1. Prove the assertion of Proposition 11.7.

These results can be extended in several directions. Together with the saddle point problem and the problem of solving a zero-sum two-person game we can consider the general case of an m-person noncooperative game. Recall that such a game consists of m players, each of which has a strategy set $X_i \subseteq \mathbb{R}^{n_i}$ and a utility function $f_i : X \to \mathbb{R}$, where

$$X = X_1 \times \ldots \times X_m.$$

A point $x^* = (u_1^*, \ldots, u_m^*)^T \in X$ is said to be a *Nash equilibrium point* for this game, if

$$\begin{array}{c} f_i(u_1^*, \ldots, u_{i-1}^*, v_i, u_{i+1}^*, \ldots, u_m^*) \leq f_i(x^*) \\ \forall v_i \in X_i, i = 1, \ldots, m; \end{array} \tag{11.20}$$

see e.g. *Aubin* (1998). Set

$$\Psi(x, y) = -\sum_{i=1}^m f_i(u_1, \ldots, u_{i-1}, v_i, u_{i+1}, \ldots, u_m), \tag{11.21}$$

$$x = (u_1, \ldots, u_m)^T, y = (v_1, \ldots, v_m)^T,$$

with $n = \sum_{i=1}^m n_i$ and

$$\Phi(x, y) = \Psi(x, y) - \Psi(x, x), \tag{11.22}$$

then the Nash equilibrium problem (11.20) becomes equivalent to the general *equilibrium problem* (EP for short) which is the problem of finding a point $x^* \in X$ such that

$$\Phi(x^*, y) \geq 0 \quad \forall y \in X. \tag{11.23}$$

Exercise 11.2. Prove the equivalence of problems (11.21) – (11.23) and (11.20).

Let us define the mapping $G : \mathbb{R}^n \to \mathbb{R}^n$ as follows:

$$G(x) = \nabla_y \Phi(x, y)\big|_{y=x} . \tag{11.24}$$

From Theorem 11.1 we also obtain the equivalence result.

Corollary 11.2. *Suppose that $\Phi(x, x) = 0$ and $\Phi(x, \cdot)$ is convex and differentiable for each $x \in X$. Then problems (11.23) and (11.1), (11.24) are equivalent.*

So, the Nash equilibrium problem can be replaced with a suitable VI.

We now give a sufficient condition for monotonicity of the mapping G defined in (11.24), which is due to *R.T. Rockafellar* and *S.P. Uryas'yev.*

Proposition 11.8. *Let $\Phi : \mathbb{R}^n \times \mathbb{R}^n \to \mathbb{R}$ be a bifunction such that $\Phi(x, x) = 0$ for every $x \in X$. Suppose that $\Phi(x, \cdot)$ and $-\Phi(\cdot, y)$ are convex and differentiable for all $x, y \in X$. If G is defined by (11.24), then it is monotone on X.*

Proof. Fix $x, y \in X$. Take any $\alpha \in (0, 1)$ and set $x_\alpha = \alpha x + (1 - \alpha)y$. It is clear that $x_\alpha \in X$ since X is convex. Due to concavity of $\Phi(\cdot, x_\alpha)$, we have

$$0 = \Phi(x_\alpha, x_\alpha) \geq \alpha\Phi(x, x_\alpha) + (1 - \alpha)\Phi(y, x_\alpha)$$

or equivalently,

$$\alpha [\Phi(x, x_\alpha) - \Phi(x, x)] \leq (1 - \alpha) [\Phi(y, y) - \Phi(y, x_\alpha)] . \tag{11.25}$$

On the other hand, by definition and Proposition 11.4, we obtain

$$\Phi(x, x_\alpha) - \Phi(x, x) \geq (x_\alpha - x)^T G(x)$$

and

$$\Phi(y, x_\alpha) - \Phi(y, y) \geq (x_\alpha - y)^T G(y).$$

Using these inequalities together with (11.25) yields

$$\begin{aligned} \alpha(x_\alpha - x)^T G(x) &\leq \alpha [\Phi(x, x_\alpha) - \Phi(x, x)] \\ &\leq (1 - \alpha) [\Phi(y, y) - \Phi(y, x_\alpha)] \\ &\leq (1 - \alpha)(y - x_\alpha)^T G(y). \end{aligned}$$

Since $x_\alpha - x = (1 - \alpha)(y - x)$ and $y - x_\alpha = \alpha(y - x)$, it follows that

$$\alpha(1 - \alpha)(y - x)^T G(x) \leq \alpha(1 - \alpha)(y - x)^T G(y),$$

or equivalently,

$$(y - x)^T [G(y) - G(x)] \geq 0.$$

Hence, G is monotone on X. □

It is clear that the saddle point problem (11.7) is a particular case of EP (11.23) if we set

$$\Phi(x, y) = L(u', v) - L(u, v'), x = (u, v)^T, y = (u', v')^T.$$

If $L : \mathbb{R}^l \times \mathbb{R}^m \to \mathbb{R}$ is (strictly, strongly) convex in u and (strictly, strongly) concave in v, then the bifunction Φ will be (strictly, strongly) concave-convex. It means that Corollary 11.1 may be deduced from Corollary 11.2. Moreover, we obtain easily the monotonicity of the mapping G in (11.8) associated with the saddle point problem.

Corollary 11.3. *If $L : \mathbb{R}^l \times \mathbb{R}^m \to \mathbb{R}$ is a differentiable convex-concave bifunction, then the mapping G in (11.8) is monotone.*

It follows that G in (11.13) is monotone, if $f_i : \mathbb{R}^l \to \mathbb{R}$, $i = 0, \dots, m$ are convex differentiable functions.

Let us now consider another generalization of the primal-dual system (11.1), (11.15), or equivalently, (11.16), where U is defined in (11.11), and V is a convex closed subset of $\mathbb{R}^m_+$. The problem is to find a pair $(u^*, v^*) \in U \times V$ such that

$$\begin{array}{c} (u - u^*)^T[F(u^*) + \sum\limits_{i=1}^{m} v_i^* \nabla f_i(u^*)] \geq 0 \quad \forall u \in U, \\ (v - v^*)^T[b(v^*) - f(u^*)] \geq 0 \quad \forall v \in V, \end{array} \tag{11.26}$$

where $b : \mathbb{R}^m_+ \to \mathbb{R}^m$ is a continuous mapping. It is equivalent to VI (11.1) with $X = U \times V$,

$$G(x) = G(u, v) = \begin{pmatrix} F(u) + \sum_{i=1}^m v_i \nabla f_i(u) \\ b(v) - f(u) \end{pmatrix}. \tag{11.27}$$

Proposition 11.9. *Suppose that $f_i : \mathbb{R}^l \to \mathbb{R}$, $i = 1, \dots, m$ are convex differentiable functions. If the mappings $F : U \to \mathbb{R}^l$ and $b : V \to \mathbb{R}^m$ are (strictly, strongly) monotone, then the mapping G in (11.27) is (strictly, strongly) monotone on $X = U \times V$.*

Proof. Choose arbitrary points $x = (u, v) \in X$ and $y = (u', v') \in X$. Then

$$\begin{aligned} &(x - y)^T[G(x) - G(y)] = (u - u')^T[F(u) - F(u')] \\ &+(u - u')^T \left[\sum_{i=1}^{m} v_i \nabla f_i(u) - \sum_{i=1}^{m} v_i' \nabla f_i(u') \right] \\ &+(v - v')^T[b(v) - b(v')] + (v - v')^T[f(u') - f(u)]. \end{aligned}$$

By Proposition 11.4, we obtain

$$(u - u')^T \nabla f_i(u) \geq f_i(u) - f_i(u')$$

and

$$(u-u')^T\nabla f_i(u') \le f_i(u) - f_i(u').$$

Therefore,

$$\begin{aligned}(x-y)^T[G(x)-G(y)] \;\ge\; & (u-u')^T[F(u)-F(u')] \\ & +(v-v')^T[b(v)-b(v')],\end{aligned}$$

and monotonicity of G follows from monotonicity of F and b. The (strict, strong) monotonicity of G can be deduced similarly. □

Note that we have also a monotonicity criterion for G in (11.15) if we set $b(v) \equiv 0$ in (11.27).

Exercise 11.3. Prove the assertion of Proposition 11.9 for the strictly and strongly monotone cases.

It follows from Proposition 11.6 that this problem also solves (11.16), (11.11) with perturbations of the right-hand side of constraints, i.e. it depends on the optimal value of the dual variables.

Proposition 11.10. *If $x^* = (u^*, v^*)$ is a solution to (11.16), (11.11) with $V = \mathbb{R}^m_+$, then u^* is a solution to the problem of finding $u^* \in \tilde{D}$ such that*

$$(u-u^*)^T F(u^*) \ge 0 \quad \forall u \in \tilde{D}, \tag{11.28}$$

$$\tilde{D} = \{u \in U \mid f_i(u) \le b_i(v^*) \quad i = 1, \dots, m\},$$

U is the same as in (11.11).

By definition, problem (11.28) belongs to so-called implicit VI's. Under certain additional assumptions, it can be converted into a saddle point problem. In fact, let us consider problem (11.7) where

$$L(u,v) = f_0(u) + \sum_{i=1}^{m} v_i f_i(u) - \varphi(v) \tag{11.29}$$

and suppose that $F(u) = \nabla f_0(u)$, $b(v) = \nabla\varphi(v)$ and that $\varphi : \mathbb{R}^m_+ \to \mathbb{R}$ and $f_i : \mathbb{R}^l \to \mathbb{R}$, $i = 0, \dots, m$ are convex differentiable functions.

Exercise 11.4. Show that the saddle point problem (11.7), (11.29) is equivalent to the system (11.26), (11.11) with $F(u) = \nabla f_0(u)$, $b(v) = \nabla\varphi(v)$ under the above assumptions.

The main motivation for studying this extended primal-dual system (11.26), (11.11) stems from the fact that it contains formulations of many equilibrium type problems described in Part I. However, this problem may be investigated as a usual VI and both the theory and solution methods of VI's are applicable for all these equilibrium problems.

11.2 Existence and uniqueness results

Most existence results of solutions for VI's are proved by using various fixed-point theorems. We will also use the famous *Brouwer* fixed-point theorem; e.g. see *Nikaido* (1968), Chapter 1, Theorem 4.3.

Proposition 11.11. *Every continuous mapping T which maps a nonempty convex and compact set X into itself has a fixed point.*

In order to apply this result to VI's we need several properties of the projection mapping. Given a point x and a set X in E, we denote by $\pi_X(x)$ the projection of x onto X:

$$\pi_X(x) \in X, \|x - \pi_X(x)\| = \min_{y \in X} \|x - y\|.$$

Proposition 11.12. *Suppose Y is a nonempty convex closed set in E, and x is an arbitrary point in E. Then:*

(i) There exists the unique projection $p = \pi_Y(x)$ of any point x onto the set Y.

(ii) A point $p \in Y$ is a projection of x onto Y if and only if

$$(p - x)^T(y - p) \geq 0 \quad \forall y \in Y. \tag{11.30}$$

(iii) The projection mapping $\pi_Y(\cdot)$ is non-expansive and

$$\begin{aligned} &(x'' - x')^T[\pi_Y(x'') - \pi_Y(x')] \\ &\geq \|\pi_Y(x'') - \pi_Y(x')\|^2 \quad \forall x', x'' \in E. \end{aligned} \tag{11.31}$$

Proof. It is clear that the point $p = \pi_Y(x)$ is a solution of the following convex optimization problem:

$$\min \to \{\varphi(y) \mid y \in Y\},$$

where $\varphi(y) = 0.5\|y - x\|^2$. Due to Theorem 11.1, this problem is equivalent to the variational inequality (11.30), i.e., assertion (ii) holds. Moreover, this problem is equivalent to the following optimization problem:

$$\min \to \{\varphi(y) \mid y \in Y, \|y - x\| < r\},$$

where

$$r > \inf_{y \in Y} \|y - x\| \geq 0;$$

which is solvable due to the well-known *Weierstrass* theorem. Take arbitrary points $x', x'' \in E$ and set $p' = \pi_Y(x')$, $p'' = \pi_Y(x'')$. Applying (11.30) with $x = x'$, $p = p'$, $y = p''$ and with $x = x''$, $p = p''$, $y = p'$, respectively, gives

$$(p' - x')^T(p'' - p') \geq 0$$

and

$$(p'' - x'')^T(p' - p'') \geq 0.$$

Adding these inequalities and applying the *Cauchy-Schwarz* inequality, we obtain

$$(p'' - p')^T(p'' - p') \leq (x'' - x')^T(p'' - p') \leq \|x'' - x'\|\|p'' - p'\|,$$

i.e., (11.31) holds, moreover,

$$\|p'' - p'\| \leq \|x'' - x'\|.$$

Therefore, the mapping $\pi_Y(\cdot)$ is non-expansive. This property yields the uniqueness of the projection, i.e. assertion (i) is also true. □

Note that assertion (iii) implies that the mapping $\pi_Y(\cdot)$ is continuous. We now obtain an equivalent fixed-point formulation of VI (11.1).

Proposition 11.13. *Let X be a nonempty, closed and convex subset of a finite - dimensional Euclidean space E. A point $x^* \in X$ solves VI (11.1) if and only if*

$$x^* = \pi_X[x^* - \theta G(x^*)] \tag{11.32}$$

for some $\theta > 0$.

Proof. If (11.32) holds, then, applying (11.30) with $p = x^*$, $x = x^* - \theta G(x^*)$, and $X = Y$, gives

$$(x^* - [x^* - \theta G(x^*)])^T(y - x^*) \geq 0 \quad \forall y \in X,$$

hence $x^* \in X^*$. Conversely, let $x^* \in X^*$, but $x^* \neq \tilde{x} = \pi_X(x^* - \theta G(x^*))$. Using (11.30) with $p = \tilde{x}$, $x = x^* - \theta G(x^*)$, $y = x^*$, and $Y = X$, we have

$$(\tilde{x} - [x^* - \theta G(x^*)])^T(x^* - \tilde{x}) \geq 0,$$

or

$$(\tilde{x} - x^*)^T[\theta G(x^*)] \leq -\|\tilde{x} - x^*\|^2 < 0,$$

i.e. $x^* \notin X^*$, a contradiction. □

We are now ready to establish an existence result for VI (11.1) by simple combining Propositions 11.11 and 11.13.

Theorem 11.2. *Let X be a nonempty, convex and compact subset of a finite - dimensional Euclidean space E and let $G : X \to E$ be a continuous mapping. Then VI (11.1) is solvable.*

For obtaining existence results on unbounded sets, we have to utilize a coercivity condition.

Theorem 11.3. *Let X be a nonempty, convex and closed subset of a finite - dimensional Euclidean space E and let $G : X \to E$ be a continuous mapping. Suppose that there exists a nonempty bounded subset Y of X such that for every $x \in X \backslash Y$ there is $y \in Y$ with*

$$(x - y)^T G(x) > 0.$$

Then VI (11.1) has a solution.

Proof. It suffices to consider only the unbounded case. Let B_r denote the closed ball (under the norm) of E with center at 0 and radius $r > 0$. Choose r large enough so that $r > \|y\|$ for every $y \in Y$. Then Theorem 11.2 guarantees the existence of a solution $x_r \in X \cap B_r$ for the following VI:

$$(z - x_r)^T G(x_r) \geq 0 \quad \forall z \in X \cap B_r.$$

Moreover, we must have $\|x_r\| < r$ due to the coercivity condition. Take an arbitrary $x \in X$. Then there exists $\varepsilon > 0$ small enough such that $x_r + \varepsilon(x - x_r) \in X \cap B_r$. It follows that

$$[x_r + \varepsilon(x - x_r) - x_r]^T G(x_r) \geq 0.$$

Dividing ε on both sides of the above inequality, we obtain

$$(x - x_r)^T G(x_r) \geq 0,$$

which shows that x_r is the solution to VI (11.1) and the result follows. □

In general, VI can have more than one solution. We now recall a simple condition which provide the uniqueness of a solution for VI (11.1).

Proposition 11.14. *If G is strictly monotone, then VI (11.1) has at most one solution.*

Proof. Suppose that there exist at least two solutions x' and x'' of VI (11.1). Then, by definition,

$$(x'' - x')^T G(x') \geq 0$$

and

$$(x' - x'')^T G(x'') \geq 0.$$

Adding these inequalities gives

$$(x'' - x')^T [G(x') - G(x'')] \geq 0.$$

Since G is strictly monotone, it follows that $x' = x''$. □

Monotonicity properties enable us to specialize coercivity conditions. For instance, the strong monotonicity provides the existence and uniqueness result.

Theorem 11.4. *Let X be a nonempty, convex and closed subset of a finite - dimensional Euclidean space E and let $G : X \to E$ be a strongly monotone and continuous mapping. Then VI (11.1) has a unique solution.*

Proof. Due to Proposition 11.14 and Theorem 11.2, it suffices to show that VI (11.1) is solvable if X is unbounded. Fix a point $\tilde{x} \in X$. Then, for each $x \in X$, we have

$$(x - \tilde{x})^T G(x) \geq (x - \tilde{x})^T G(\tilde{x}) + \tau \|x - \tilde{x}\|^2 \to +\infty$$

as $\|x - \tilde{x}\|^2 \to +\infty$. Therefore, the coercivity condition in Theorem 11.3 holds, i.e. VI (11.1) is solvable. □

In view of Theorem 11.1 and Proposition 11.3, the above properties yield similar existence and uniqueness results for the optimization problem (11.5) where f is differentiable and strictly (strongly) convex. However, these results remain valid without differentiability.

Proposition 11.15. *Let X be a nonempty, convex and closed subset of a finite - dimensional Euclidean space E.*

(i) If $f : X \to \mathbb{R}$ is strictly convex, then (11.5) has at most one solution.

(ii) If $f : X \to \mathbb{R}$ is strongly convex and continuous, then (11.5) has a unique solution.

Proof. In case (i), suppose that there exist at least two distinct solutions x' and x'' of (11.5). Then $f(x') = f(x'') = f^*$, and, by setting $\tilde{x} = 0.5(x' + x'')$ and using the strict convexity, we have

$$f(\tilde{x}) < 0.5f(x') + 0.5f(x'') = f^*,$$

which is a contradiction.

In case (ii), it suffices to show that (11.5) is solvable in the case when X is unbounded because of (i). Fix $z \in X$ and take an arbitrary sequence

$\{x^k\} \subset X$ such that $\|x^k\| \to \infty$ as $k \to \infty$. Set $z^k = \alpha_k x^k + (1-\alpha_k)z$ with $\alpha_k = \|x^k - z\|^{-1}$, then $\|z^k - z\| = 1$ and $f(z^k) - f(z) \geq \gamma > -\infty$ because of continuity of f. Since f is strongly convex, we have

$$f(z^k) \leq \alpha_k f(x^k) + (1-\alpha_k)f(z) - 0.5\alpha_k(1-\alpha_k)\tau\|x^k - z\|^2;$$

hence

$$\begin{aligned} f(x^k) - f(z) &\geq \alpha_k^{-1}[f(z^k) - f(z)] + 0.5(1-\alpha_k)\tau\alpha_k^{-2} \\ &\geq \alpha_k^{-1}\gamma + 0.5(1-\alpha_k)\tau\alpha_k^{-2} \to +\infty \end{aligned}$$

as $k \to \infty$. It follows that the level set

$$L(z) = \{x \in X \mid f(x) \leq f(z)\}$$

is bounded. The existence of a solution of (11.5) follows from the Weierstrass theorem. This proves assertion (ii). □

Chapter 12

Applications

In this chapter, we consider applications of the theory of variational inequalities from the previous chapter to equilibrium models described in Part I.

12.1 Cassel-Wald equilibrium models

The model was described in Section 5.1. Recall that the corresponding economic system deals in n commodities and m pure factors of production. Let c_k denote the price of the k-th commodity, b_i the total inventory of the i-th factor, and a_{ij} the consumption rate of the i-th factor which is required for producing one unit of the j-th commodity. Also, let x_j denote the output of the j-th commodity. Set $c = (c_1, \ldots, c_n)^T$, $x = (x_1, \ldots, x_n)^T$, $b = (b_1, \ldots, b_m)^T$, and $A = (a_{ij})_{m \times n}$ and assume that prices are dependent of outputs, i.e. there exists a mapping $c : \mathbb{R}^n_+ \to \mathbb{R}^n_+$. Then (see (5.2)), the equilibrium point solves VI: Find $x^* \in D$ such that

$$(x^* - x)^T c(x^*) \geq 0 \quad \forall x \in D, \tag{12.1}$$

where

$$D = \{x \in \mathbb{R}^n \mid Ax \leq b, x \geq 0\}.$$

It means that the optimal outputs yield the maximal value of income subject to resource constraints when the prices are fixed at these outputs. Clearly, (12.1) is a particular case of VI (11.1). Moreover, the set D is clearly convex and closed. Also, it is nonempty if A and b contain only non-negative entries.

Exercise 12.1. Prove that D is bounded if A and b contain only non-negative entries and there is no zero column in A.

Then D is nonempty, convex, and compact, and, due to Theorem 11.1, VI (12.1) is solvable if c is continuous.

Also, we can write the optimality condition for (12.1) in the form of a system of primal-dual VIs: Find $x^* \geq 0$ and $p^* \geq 0$ such that

$$\begin{array}{c} (x^* - x)^T c(x^*) + (Ax - Ax^*)^T p^* \geq 0 \quad \forall x \geq 0; \\ (p - p^*)^T (b - Ax^*) \geq 0 \quad \forall p \geq 0; \end{array} \tag{12.2}$$

or equivalently, in the form of a system of CPs (see (5.1)). Here $p = (p_1, \ldots, p_m)^T$ is the vector of factors prices. Due to Proposition 11.6, we conclude that solvability of (12.1) implies the solvability of (12.2), moreover, p^* then gives the equilibrium factor prices.

It is rather natural to suppose that increase of outputs (supply) leads to decrease of prices. It means that the mapping $-c$ will be monotone in general and even its strengthened monotonicity properties may be derived. As to the primal-dual VI (12.2), its cost mapping then may be only monotone because b is constant; see Proposition 11.9. Thus, the general theory appears useful for the Cassel-Wald models.

12.2 Walrasian equilibrium models and their modifications

Let us now consider economic models based on equilibrium relations between supply and demand depending on prices. Such models were suggested by *L. Walras* and are usually formulated as CP; see (5.3). More precisely, the economy with perfect competition deals in n commodities. Given a price vector $p = (p_1, \ldots, p_n)^T \in \mathbb{R}^n_+$, the producers (sectors) determine their total supply $S(p) \in \mathbb{R}^n$, whereas the consumers determine their total demand $D(p) \in \mathbb{R}^n$. For simplicity, both the mappings S and D are supposed to be single-valued. Then we can define the excess demand

$$E(p) = D(p) - S(p).$$

Recall that a vector p^* is said to be an *equilibrium price* if it satisfies the following conditions:

$$p^* \in \mathbb{R}^n_+, \quad -E(p^*) \in \mathbb{R}^n_+, \quad (p^*)^T E(p^*) = 0; \tag{12.3}$$

which obviously coincide with the usual CP with the cost mapping $G = -E$ and the basic cone $X = \mathbb{R}^n_+$; see (11.2). Due to Proposition 7.1, it is equivalent to the following VI: Find $p^* \in \mathbb{R}^n_+$ such that

$$(p^* - p)^T E(p^*) \geq 0 \quad \forall p \in \mathbb{R}^n_+. \tag{12.4}$$

Observe that many popular Walrasian type equilibrium models utilize order monotonicity properties, some of them were presented in Part II. Moreover, their excess demand mapping does not possess the continuity properties over the set $\mathbb{R}^n_+$; this causes the necessity of developing certain special techniques for their substantiation; see e.g. *Arrow and Hahn* (1971), *Morishima* (1964), and *Nikaido* (1968). Nevertheless, we intend to first present examples of some other models that admit the techniques presented in Chapter 11.

Let us consider the supply mapping. Usually, the supply of a separate producer is represented as a solution set of the following optimization problem:

$$\max \rightarrow p^T y \tag{12.5}$$

subject to

$$y \in Y,$$

where Y is the technology set, which is assumed to be nonempty, convex, and compact. Then problem (12.5) is solvable, but we introduce an additional condition, which ensures the uniqueness of its solution, for the sake of simplicity of exposition.

The set Y is said to be *strictly convex*, if for each pair of points $x, y \in Y$ and for each number $\alpha \in (0,1)$, the point $\alpha x + (1-\alpha)y$ lies in $\operatorname{int} Y$.

Proposition 12.1. *If the set Y is nonempty, strictly convex and compact and $p \neq 0$, then (12.5) has a unique solution.*

Exercise 12.2. Prove the assertion of Proposition 12.1.

Moreover, then the supply mapping possesses other useful properties.

Proposition 12.2. *If the set Y is nonempty, strictly convex and compact, then the supply mapping $p \mapsto y(p)$ is continuous, positively homogeneous of degree 0, and monotone on $\mathbb{R}^n_+ \setminus \{0\}$.*

Exercise 12.3. Prove the assertion of Proposition 12.2.

Observe that the mapping $p \mapsto y(p)$ maintains in fact certain basic useful properties at 0, but it simply becomes multi-valued. Since all the properties given in Propositions 12.1 and 12.2 are additive, they remain the same for the arbitrary number of producers, hence the total supply mapping $p \mapsto S(p)$ will possess these properties.

We now consider similar properties of demand. Each consumer's demand may be represented as a mapping $p \mapsto x(p)$, where $x(p)$ solves the optimization problem

$$\max \rightarrow \varphi(x) \tag{12.6}$$

subject to

$$p^T x \leq \omega, x \in X;$$

where φ is the utility function, X is a convex cone representing the set of possible collections of commodities, and $\omega > 0$ is the value of his/her budget, which is assumed to be fixed. Then we have a useful representation of the demand mapping; its substantiation is taken from the book by *Polterovich* (1990), Chapter 3, Lemma 1.

Lemma 12.1. *Suppose that*

(a) $\varphi : X \to \mathbb{R}$ *is positively homogeneous of degree* $\alpha > 0$*;*

(b) there is $x \in X$ *such that* $\varphi(x) > 0$.

Then, the point $x(p)$ *solves problem (12.6) if and only if it solves the problem*

$$\max \to \frac{\omega}{\alpha} \ln \varphi(x) - p^T x \tag{12.7}$$

subject to

$$x \in X.$$

Proof. From condition (a) it follows that $p^T x(p) = \omega$. Suppose that there exists a point $\tilde{x} \in X$ such that $f(\tilde{x}) > f(x(p))$, where $f(x) = \frac{\omega}{\alpha} \ln \varphi(x) - p^T x$. Then $\varphi(\tilde{x}) > 0$, hence $p^T \tilde{x} > 0$, otherwise the cost function in (12.6) is not bounded above. Set $y(\tau) = \tau x(p)$ for $\tau > 0$, then the function

$$\begin{aligned} f(y(\tau)) &= \frac{\omega}{\alpha} \ln [\tau^\alpha \varphi(x(p))] - \tau p^T x(p) \\ &= \omega \ln \tau + \frac{\omega}{\alpha} \ln \varphi(x(p)) - \tau \omega \end{aligned}$$

attains its maximum at $\tau = 1$. Hence,

$$f(\tilde{x}) > f\left[(\omega^{-1} p^T \tilde{x}) x(p)\right],$$

which is equivalent to

$$\varphi\left[\omega (p^T \tilde{x})^{-1} \tilde{x}\right] > \varphi(x(p)),$$

thus contradicting the definition of $x(p)$, since the point $y = \omega(p^T \tilde{x})^{-1} \tilde{x}$ satisfies all the constraints of problem (12.6). Conversely, let $\tilde{x}$ solve problem (12.7). If $p^T \tilde{x} \leq 0$, then each point $\tau \tilde{x}$ is feasible if $\tau > 0$ and

$$\begin{aligned} f(\tau \tilde{x}) &= \frac{\omega}{\alpha} \ln [\tau^\alpha \varphi(\tilde{x})] - \tau p^T \tilde{x} \\ &= \omega \ln \tau + \frac{\omega}{\alpha} \ln \varphi(\tilde{x}) - \tau p^T \tilde{x} \to +\infty \end{aligned}$$

as $\tau \to +\infty$, thus contradicting the fact that it attains its maximum at $\tau = 1$. Hence, $p^T\tilde{x} > 0$, but the above expression attains its unique maximum at $\tau = \omega / p^T\tilde{x}$ and we obtain $p^T\tilde{x} = \omega$. Take any x such that $p^T x \leq \omega$, $x \in X$. Then

$$f(\tilde{x}) = \frac{\omega}{\alpha}\ln\varphi(\tilde{x}) - \omega \geq f(x) \geq \frac{\omega}{\alpha}\ln\varphi(x) - \omega.$$

If $\varphi(x) > 0$, then $\varphi(\tilde{x}) \geq \varphi(x)$. This means that $\tilde{x}$ is a solution of (12.6). Therefore, the assertion is true. □

This property yields the monotonicity of the mapping $p \mapsto -x(p)$.

Proposition 12.3. *If the assumptions of Lemma 12.1 are fulfilled and φ is strictly convex, then the mapping $p \mapsto -x(p)$ defined by (12.6) is single-valued and monotone.*

Proof. Note that strict convexity of φ ensures the uniqueness of a solution for (12.6). Take any $p', p'' \in \mathbb{R}^n_+$ and set $x' = x(p'), x'' = x(p'')$. Then, by Lemma 12.1, we have

$$\frac{\omega}{\alpha}\ln\varphi(x') - (p')^T x' \geq \frac{\omega}{\alpha}\ln\varphi(x'') - (p')^T x''$$

and

$$\frac{\omega}{\alpha}\ln\varphi(x'') - (p'')^T x'' \geq \frac{\omega}{\alpha}\ln\varphi(x') - (p'')^T x'.$$

Adding these inequalities yields

$$(p'' - p')^T(x' - x'') \geq 0,$$

i.e., the mapping $p \mapsto -x(p)$ is monotone. □

Exercise 12.4. Prove that the mapping $p \mapsto x(p)$ is continuous on $\mathbb{R}^n_+ \setminus \{0\}$.

Again, since these properties are additive, they are the same for each finite number of consumers and the total demand mapping $p \mapsto D(p)$ will also possess these properties. Therefore, the mapping $p \mapsto -E(p)$ will be continuous and monotone on $\mathbb{R}^n_+ \setminus \{0\}$. In general, it is sufficient for applying the results of Chapter 11 to this equilibrium model. However, in order to provide the correctness of such a result, we should somewhat modify the basic problem. In fact, instead of CP (12.3) or VI (12.4) we then introduce the slightly modified VI: Find $p^* \in P$ such that

$$(p^* - p)^T E(p^*) \geq 0 \quad \forall p \in P, \tag{12.8}$$

where

$$P = \{p \in \mathbb{R}^n \mid p_i \geq p'_i > 0 \quad \text{for } i = 1, \dots, n\}.$$

Thus, we obtain an equilibrium model with lower positive restrictions on prices; see also Sections 8.2 and 10.1. Then the cost mapping becomes well-defined, monotone and continuous over the feasible set.

There exist some other models of supply and demand which differ from those in (12.5) and (12.6). For instance, another demand mapping is presented in extended Scarf's equilibrium model described in Section 5.2; see (5.5) and (5.6). The equilibrium in this model is represented by a pair (p^*, x^*) such that

$$\begin{array}{c} p^* \geq 0, r + A^T x^* - D(p^*) \geq 0, \\ (p^*)^T [r + A^T x^* - D(p^*)] = 0; \\ x^* \geq 0, c(x^*) - Ap^* \geq 0, (x^*)^T [c(x^*) - Ap^*] = 0; \end{array} \tag{12.9}$$

where $x = (x_1, \dots, x_l)^T$ is the activity level vector of l producers, the j-th row a^j of the $l \times n$ technology matrix A gives the output of the j-th producer with the unit activity level, $c(x) = (c_1(x_1), \dots, c_l(x_l))^T$ is the marginal cost vector of producers, and $r = (r_1, \dots, r_n)^T \geq 0$ is the total endowment vector of consumers. It means that the supply is defined by

$$S(p) = A^T x + r,$$

and that the third row in (12.9) gives the optimal levels of activity under the prices p^*, whereas the first and second rows in (12.9) represent a specialization of the equilibrium conditions (12.3). Since (12.9) is a system of CPs, it can be equivalently rewritten as the system of VIs: Find a pair $(p^*, x^*) \in \mathbb{R}^n_+ \times \mathbb{R}^l_+$ such that

$$\begin{array}{r} (p - p^*)^T [r - D(p^*) + A^T x^*] \geq 0 \quad \forall p \in \mathbb{R}^n_+, \\ (x - x^*)^T [c(x^*) - Ap^*] \geq 0, \quad \forall x \in \mathbb{R}^l_+; \end{array} \tag{12.10}$$

which is clearly a particular case of the extended primal-dual VI (11.26); see also (4.11). Note that the case of zero marginal cost used in the initial model also leads to positive homogeneity of degree 0 of supply.

By construction, the marginal cost mapping $c(x)$ is monotone and diagonal, i.e. there exists a continuous convex function

$$f(x) = \sum_{j=1}^{l} f_j(x_j) \quad \text{with} \quad f'_j(x_j) = c_j(x_j);$$

see also (5.4). Next, if we suppose that behavior of consumers is represented by the model (12.6) with fixed budgets, then, by Proposition 12.3, the mapping $-D(p)$ is monotone and in fact the gradientmap of a convex function

$\psi(p)$. By Proposition 11.9, system (12.10) is then a monotone VI, and it is also equivalent to the saddle point problem: Find $(p^*, x^*) \in \mathbb{R}^n_+ \times \mathbb{R}^l_+$ such that

$$L(p^*, x) \leq L(p^*, x^*) \leq L(p, x^*) \quad \forall p \in \mathbb{R}^n_+, \forall x \in \mathbb{R}^l_+.$$

where

$$L(p, x) = r^T p + \psi(p) + Ax - f(x),$$

which is convex-concave. If the model involves additional restrictions on prices and activity levels (see (12.8)), then they can be handled easily by replacing $\mathbb{R}^n_+$ and $\mathbb{R}^l_+$ with the corresponding feasible sets $P \subseteq \mathbb{R}^n_+$ and $X \subseteq \mathbb{R}^l_+$. Thus, most results of the general theory can be applied to this model.

It has been mentioned that the extended Scarf model is written as the primal-dual system (4.11) or (4.12), which is a particular case of the system (11.26) with the affine constraint functions. Therefore, all the models formulated as system (4.11): Find a pair (x^*, y^*) such that

$$\begin{array}{ll} x^* \geq 0, & y^* \geq 0; \\ A^T y^* - c(x^*) \geq 0, & b(y^*) - Ax^* \geq 0; \\ (x^*)^T[A^T y^* - c(x^*)] = 0, & (y^*)^T[b(y^*) - Ax^*] = 0; \end{array} \tag{12.11}$$

can be investigated and solved by using the results of Chapter 11. Several examples of such models are given in Part I. They include all the spatial price equilibrium models from Section 5.3 and the multicommodity formulation of the network equilibrium model from Section 6.1.

Let us now consider some other economic equilibrium model which seems intermediate between Cassel-Wald and Scarf ones; such kinds of models can be found in the books by *Ahn* (1979) and by *Arrow, Hurwicz, and Uzawa* (1958), Chapter 15. The model describes an n-commodity market and includes consumption and production with linear technology. The consumption sector is described by its demand mapping $d : \mathbb{R}^n_+ \to \mathbb{R}^n$. The production sector is described with the help of the $m \times l$ industry consumption matrix A and $n \times l$ industry output matrix Q; i.e. it is similar to that in the model from Section 3.2. It means that, given an activity levels vector of technologies $x = (x_1, \ldots, x_l)^T$, the resource consumption is Ax and the output is Qx. Moreover, if $c = (c_1, \ldots, c_l)^T$ is the costs of unit level technologies, the total cost is given by $c^T x$. Therefore, the model can be rewritten as the implicit optimization problem:

$$\min \to c^T x \tag{12.12}$$

subject to

$$Ax \leq b,$$

$$Qx \geq d(p^*).$$
$$x \geq 0.$$

where $b \in \mathbb{R}^m$ is the vector of resource endowments, $p^* \in \mathbb{R}^n_+$ is the (unknown) price equilibrium vector. This means that the production sector seeks to the optimal plan of activity levels, which minimizes pure production expenses subject to resource constraints so that the supply values can not be less than the equilibrium demand ones. To make the sense of this problem more explicit, we fix p^* temporarily, i.e. set $d = d(p^*)$. Then (12.12) becomes a linear programming problem. By using the optimality conditions for this problem (see Theorems 4.3 and 4.4), we see that (12.12) is then equivalent to the problem of finding $(x^*, u^*, v^*) \in \mathbb{R}^n \times \mathbb{R}^m \times \mathbb{R}^m$ such that

$$\begin{aligned} &x^* \geq 0, c + A^T u^* - Q^T v^* \geq 0, (x^*)^T (c + A^T u^* - Q^T v^*) = 0; \\ &u^* \geq 0, b - Ax^* \geq 0, (u^*)^T (b - Ax^*) = 0; \\ &v^* \geq 0, Qx^* - d \geq 0, (v^*)^T (Qx^* - d) = 0; \end{aligned} \qquad (12.13)$$

with adding the equilibrium condition $p^* = v^*$ and $d = d(p^*)$. If we set $\tilde{c} = c - Q^T v^*$, the same optimality conditions imply that the first and second rows in (12.13) are equivalent to the optimization problem

$$\min \to \tilde{c}^T x$$

subject to

$$Ax \leq b, x \geq 0;$$

or equivalently,

$$(x - x^*)^T (c - Q^T v^*) \geq 0 \quad \forall x \in D,$$

where

$$D = \left\{ x \in \mathbb{R}^n_+ \mid Ax \leq b \right\}. \qquad (12.14)$$

In this problem, the production sector simply minimizes its total loss subject to the resource constraints. Moreover, since $Qx^* = S(p^*)$ is the industrial demand, the third row in (12.13) is nothing but the usual equilibrium conditions (12.3). Combining all these properties, we conclude that the initial model is written as the system of VI's: Find a pair $(p^*, x^*) \in \mathbb{R}^n_+ \times D$ such that

$$\begin{aligned} (p - p^*)^T (Qx^* - d(p^*)) \geq 0 \quad &\forall p \in \mathbb{R}^n_+, \\ (x - x^*)^T (c - Q^T p^*) \geq 0 \quad &\forall x \in D; \end{aligned} \qquad (12.15)$$

which is also a particular case of the system (11.26).

12.3 Existence results in Walrasian equilibrium models

We now return to the problem of establishing existence results in economic equilibrium models formulated as CP (12.3) in the case when the excess demand mapping E does not possess continuity properties on $\mathbb{R}^n_+$. In several previous models, E is not defined at 0, in some other models, described in Chapters 8 and 10, it is not defined on the boundary of $\mathbb{R}^n_+$. Then we can not apply directly the results of Chapter 11. However, under certain natural assumptions, these results may be adjusted to CP (12.3) without continuity on $\mathbb{R}^n_+$.

Our considerations are based on the following simple property.

Lemma 12.2. *Suppose that*

(a) $E : \mathbb{R}^n_+ \setminus \{0\} \to \mathbb{R}^n$ is positively homogeneous of degree 0, i.e. $E(\lambda p) = E(p)$ for all $\lambda > 0$ and $p \in \mathbb{R}^n_+ \setminus \{0\}$.

(b) The Walras law holds, i.e.

$$p^T E(p) = 0 \quad \forall p \in \mathbb{R}^n_+ \setminus \{0\}.$$

Then CP (12.3) is equivalent to the following VI: Find $p^* \in Q$ such that

$$(p^* - p)^T E(p^*) \geq 0 \quad \forall p \in Q, \tag{12.16}$$

where

$$Q = \left\{ p \in \mathbb{R}^n_+ \;\middle|\; \sum_{i=1}^{n} p_i = 1 \right\}$$

in the sense that $Q^* = P^* \bigcap Q$, where P^* and Q^* denote the solution sets of CP (12.3) and VI (12.16), respectively.

Proof. Let $p^* \neq 0$ solve CP (12.3), then $\tilde{p} = (\sum_{i=1}^{n} p_i^*)^{-1} p^* \in Q$ and is also a solution of CP (12.3), hence it solves VI (12.4) due to Proposition 7.1. We see that $\tilde{p}$ also solves (12.16) and $P^* \bigcap Q \subseteq Q^*$. Conversely, if p^* solves (12.16), then (b) yields

$$p^T E(p^*) \leq 0 \quad \forall p \in Q.$$

Denote by e^i the i-th coordinate vector in $\mathbb{R}^n$, i.e.

$$e^i_j = \begin{cases} 1 & \text{if } i = j, \\ 0 & \text{if } i \neq j. \end{cases}$$

Then, setting p to be e^i, we obtain $E_i(p^*) \leq 0$ for $i = 1, \dots, n$. Therefore p^* solves CP (12.3) and $Q^* \subseteq P^* \bigcap Q$. □

Observe that both conditions (a) and (b) are usual for most Walrasian equilibrium models presented in Parts I and II.

Theorem 12.1. *Suppose that the excess demand mapping* $E : \mathbb{R}^n_+ \setminus \{0\} \rightarrow \mathbb{R}^n$ *is continuous on* $\mathbb{R}^n_+ \setminus \{0\}$, *positive homogeneous of degree* 0, *and satisfies the Walras law. Then CP (12.3) has a solution.*

Proof. From Theorem 11.2 it follows that VI (12.16) is solvable, but Lemma 12.2 shows that each its solution also solves CP (12.3), as desired. □

We now consider the case when E is undefined on the boundary of $\mathbb{R}^n_+$.

Theorem 12.2. *Suppose that the excess demand mapping* $E : \mathbb{R}^n_+ \rightarrow \mathbb{R}^n$ *is continuous and positive homogeneous of degree* 0 *on* $\mathrm{int}\mathbb{R}^n_+$, *and the Walras law holds on* $\mathrm{int}\mathbb{R}^n_+$. *Suppose also that, for any sequence* $\{p^k\} \subset \mathrm{int}\mathbb{R}^n_+$ *converging to* p, *it holds that*

$$E_i(p^k) \begin{cases} \rightarrow +\infty & \text{if } p_i = 0, \\ \geq C > -\infty & \text{if } p_i > 0, \end{cases}$$

as $k \rightarrow \infty$. *Then CP (12.3) has a solution, which also solves the problem*

$$E(p^*) = 0, \ p^* > 0. \tag{12.17}$$

Proof. For any $\rho > 0$, set

$$Q_\rho = \{p \in Q \mid p_i \geq \rho \quad \text{for } i = 1, \ldots, n\}.$$

Then E is continuous on Q_ρ and the problem of finding $p(\rho) \in Q_\rho$ such that

$$(p(\rho) - p)^T E(p(\rho)) \geq 0 \quad \forall p \in Q_\rho \tag{12.18}$$

has a solution due to Theorem 11.2. We proceed to show that there exists $\rho' > 0$ small enough such that $p_i(\rho') > \rho'$ for $i = 1, \ldots, n$. On the contrary, suppose that $\{q^k\} \rightarrow q' \in Q \setminus \mathrm{int}\mathbb{R}^n_+$ where $q^k = p(\rho_k)$ for some sequence $\{\rho_k\} \rightarrow 0$. Then we take $p = (1, \ldots, 1)^T$ and (12.18) together with the Walras law yield

$$\sum_{i=1}^{n} E_i(q^k) \leq 0,$$

a contradiction. Set $p^* = p(\rho')$ and choose an arbitrary point $p \in Q$. If $p \in Q_{\rho'}$, then (12.16) holds. If $p \notin Q_{\rho'}$, there exists $\mu \in (0, 1)$ such that $\mu p + (1 - \mu)p^* \in Q_{\rho'}$ and (12.18) now gives

$$[p^* - \mu p - (1 - \mu)p^*]^T E(p^*) \geq 0,$$

i.e.

$$(p^* - p)^T E(p^*) \geq 0.$$

It means that p^* solves VI (12.16), hence it solves CP (12.3), which is proved along the lines of the second part of Lemma 12.2. However, $p^* \in \mathrm{int}\mathbb{R}^n_+$ and (12.3) now implies (12.17). □

Observe that many price equilibrium models described in Sections 8.2, 8.3, and 10.1, satisfy the above conditions.

12.4 Imperfect competition models

This model given in Section 5.4 describes an oligopolistic market involving n firms supplying a homogeneous product and represents a non-cooperative game, where the i-th player has the strategy set $\mathbb{R}_+$ and the utility function

$$f_i(x) = x_i p(\sigma_x) - h_i(x_i), \tag{12.19}$$

where $x = (x_1, \dots, x_n)$, x_i is the supply of the i-th firm, $h_i : \mathbb{R}_+ \to \mathbb{R}$ is the cost function, $p : \mathbb{R}_+ \to \mathbb{R}$ is the inverse demand (or price) function of the market, and $\sigma_x = \sum\limits_{i=1}^{n} x_i$ is the total supply in the market.

This problem is reduced to the equivalent VI under rather general assumptions; see Theorem 10.2 and Corollary 11.2. This is the case if the i-th firm profit function f_i is concave in x_i for $i = 1, \dots, n$. In order to satisfy this condition we can suppose that the cost functions h_i are convex, the inverse demand function p is non-increasing, and that the industry revenue function $\mu(\sigma) = \sigma p(\sigma)$ is concave on $\mathbb{R}_+$; see Lemma 10.1. Under the corresponding differentiability conditions on the functions p and h_i, the equivalent CP (9.2) (or VI) has the cost mapping $G : \mathbb{R}^n_+ \to \mathbb{R}^n$ defined by

$$G_i(x) = h'_i(x_i) - p(\sigma_x) - x_i p'(\sigma_x), \quad i = 1, \dots, n; \tag{12.20}$$

see (10.9). In Section 10.2, various existence and uniqueness results for the oligopolistic equilibrium problem were obtained by using order monotonicity properties of the mapping G. They remain valid if there exist lower and upper bounds for output levels, then each i-th player strategy set is the segment $[\alpha_i, \beta_i]$ with $0 \leq \alpha_i < \beta_i \leq +\infty$ for $i = 1, \dots, n$ and, by Corollary 11.2, the equilibrium problem becomes equivalent to VI (11.1), (12.20), where the feasible set X is given in (7.4), or to MCP (see Proposition 7.2). However, the applied problem may contain additional restrictions on outputs, which leads to more general VI (11.1), (12.20), where the order monotonicity properties are not sufficient in general. For this reason, we intend to investigate the usual monotonicity properties of the mapping G

in (12.20). By definition, G is also defined by formula (11.24) where the bifunction Φ is defined by (11.21), (11.22), and (12.19). Therefore, we can deduce the monotonicity of G from the assertion of Proposition 11.8. It suffices to show that the bifunction

$$\Phi(x,y) = \sum_{i=1}^{n} [f_i(x) - f_i(x_1,\dots,x_{i-1},y_i,x_{i+1},\dots,x_n)] \tag{12.21}$$

is concave-convex. Clearly, $\Phi(x,\cdot)$ is convex since each function $f_i(x)$ is concave in x_i. The concavity of $\Phi(\cdot,y)$ requires an additional assumption.

Proposition 12.4. *Under the assumptions above and the convexity of* $p : \mathbb{R}_+ \to \mathbb{R}$*, the function* $\Phi(\cdot,y)$ *in (12.21) is concave for each* $y \in \mathbb{R}^n_+$.

This property can be deduced from the consideration of the separate functions in (12.21), (12.19).

Exercise 12.5. Prove the assertion of Proposition 12.4.

Corollary 12.1. *Under the above assumptions and the convexity of* $p : \mathbb{R}_+ \to \mathbb{R}$*, the mapping* G *in (12.20) is monotone.*

The result follows from Propositions 12.4 and 11.8.

In Section 5.4, the general equilibrium model for auction markets was described and formulated as VI (5.28), (5.29). Hence, its analysis may be based upon properties presented in Chapter 11.

Exercise 12.6. Establish monotonicity criteria of the cost mapping and find existence and uniqueness results for VI (5.28), (5.29).

12.5 Network and migration equilibrium models

We now discuss applicability of the results of Chapter 11 to the models described in Chapter 6. Although the multicommodity formulation of the network equilibrium model is a particular case of the primal-dual system (12.11), the path flow formulation (6.8) is a special case of VI (11.1). By definition, its solution $x^* \in X$ satisfies the variational inequality:

$$\sum_{l \in A} F_l(f^*)(f_l - f_l^*) \geq 0, \ f = Bx, \quad \forall x \in X, \tag{12.22}$$

where $f^* = Bx^*$,

$$\begin{array}{c} X = \prod_{w \in W} X_w, \\ X_w = \left\{ x \,\middle|\, \sum_{p \in P_w} x_p = d_w, x_p \geq 0 \ p \in P_w \right\}, \\ w \in W; \end{array} \tag{12.23}$$

$f = (f_l)_{l \in A}$ is the vector of flow costs (delays), $x = (x_p)_{p \in P_w, w \in W}$ is the vector of path flows, B is the arc-path incidence matrix, A is the set of arcs in the transportation network, W is the set of origin-destination pairs of nodes, P_w is the set of paths joining pair w, $d_w \geq 0$ is the traffic demand for this pair. Also, F_l determines the cost for arc l, which depends on the distribution of flows. Note that (12.22) may be rewritten as follows:

$$\begin{aligned} &\sum_{w \in W} \sum_{p \in P_w} G_p(x^*)(x_p - x_p^*) \\ &= (x - x^*)^T [B^T F(Bx^*)] \geq 0 \quad \forall x \in X. \end{aligned} \tag{12.24}$$

Clearly, the feasible set X in (12.23) is nonempty, convex, and compact. If F is continuous, then problem (12.22), (12.23) is solvable due to Theorem 11.2.

Exercise 12.7. Prove that the monotonicity of F implies the monotonicity of G.

It should be noted that a similar assertion on strict (strong) monotonicity is not true.

The migration equilibrium model (6.9), (6.12) is also a special case of VI (11.1). Moreover, its feasible set is clearly nonempty, convex and closed, but it is also bounded; see Exercise 6.2. Then the existence result for this problem may be deduced from Theorem 11.2 under only continuity of the utility and migration cost mappings.

We now show that equivalence between (6.9), (6.12) and (6.9)–(6.11) follows from Proposition 11.7. For the sake of convenience, we rewrite (6.9), (6.12) here. It is the problem of finding a pair $(\mathbf{x}^*, \mathbf{h}^*) \in H$ such that

$$\begin{aligned} &\sum_{i \in \mathbf{N}} (x_i^* - x_i) u_i(\mathbf{x}^*) \\ &+ \sum_{i,j \in \mathbf{N},\ i \neq j} (h_{ij} - h_{ij}^*) c_{ij}(\mathbf{h}^*) \geq 0 \quad \forall (\mathbf{x}, \mathbf{h}) \in H, \end{aligned}$$

where

$$\begin{aligned} H = \ &\Bigg\{ (\mathbf{x}, \mathbf{h}) \ \Bigg| \ \mathbf{h} \geq 0,\ \sum_{j \neq i} h_{ij} \leq b_i, \\ &\qquad x_i = b_i + \sum_{j \neq i} h_{ji} - \sum_{j \neq i} h_{ij},\ \forall i \in \mathbf{N} \Bigg\}, \end{aligned}$$

$\mathbf{x} = (x_i \mid i \in \mathbf{N})$, $\mathbf{h} = (h_{ij} \mid i, j \in \mathbf{N}, i \neq j)$, and $\mathbf{N}$ is the set of nodes. From Proposition 11.7 we obtain the following necessary and sufficient optimality conditions for the VI: Find $(\mathbf{x}^*, \mathbf{h}^*, \lambda, \mu)$ such that

$$\begin{cases} -u_i(\mathbf{x}^*) - \lambda_i = 0 \quad i \in \mathbf{N}; \\ h_{ij}^* \geq 0, c_{ij}(\mathbf{h}^*) - \lambda_i + \lambda_j + \mu_i \geq 0, \\ h_{ij}^*(c_{ij}(\mathbf{h}^*) - \lambda_i + \lambda_j + \mu_i) = 0 \ i, j \in \mathbf{N},\ i \neq j; \\ b_i + \sum\limits_{j \neq i} h_{ji}^* - \sum\limits_{j \neq i} h_{ij}^* = 0 \quad i \in \mathbf{N}; \\ \mu_i \geq 0, \sum\limits_{j \neq i} h_{ij}^* - b_i \leq 0, \mu_i(\sum\limits_{j \neq i} h_{ij}^* - b_i) = 0 \quad i \in \mathbf{N}. \end{cases}$$

Setting $\lambda_i = -u_i(\mathbf{x}^*)$ in the second and third rows gives (6.10), whereas the fourth and fifth rows give (6.9) and (6.11), respectively. Therefore, the migration equilibrium conditions are represented by VI.

The monotonicity properties of the negative utility and migration cost mappings seem very natural, but they imply the monotonicity of the problem (6.9), (6.12). Therefore, most results of the theory of VIs can be applied to these equilibrium problems.

Chapter 13

Projection Type Methods

The role of iterative methods for equilibrium problems is twofold. First of all, they are considered as a basis for construction of efficient computational procedures of finding equilibrium points. However, they can also be treated as models of dynamic processes in the systems under consideration, hence, their convergence properties then describe stability of these systems and enable us to evaluate efficiency of the control procedures. For this reason, investigation of properties of some methods, which may be inefficient from the computational point of view, but seem very natural dynamic processes, may give non-trivial conclusions about behavior of real systems. In this chapter, we consider the projection method and its extensions for VI (11.1), which is very popular due to its simplicity, clarity, and likeness to natural dynamical processes such as the Walras tâtonnement (see Section 10.1). At the same time, its convergence properties serve as a basis for understanding more complicated methods. The detailed description of various solution methods for VIs can be found in the books by *Facchinei and Pang* (2003), *Konnov* (2001), and *Patriksson* (1999).

13.1 The classical projection method

Let us consider VI (11.1) where X is a nonempty, closed and convex subset of a finite-dimensional Euclidean space E and $G : X \to E$ is a continuous mapping. As before, X^* denotes the solution set of this problem. In the usual *projection method*, the iteration sequence is constructed in conformity with the rule:

$$x^{k+1} = \pi_X[x^k - \lambda_k G(x^k)], \ \lambda_k > 0, \tag{13.1}$$

where x^k is the current iteration point, $x^0 \in X$, and $\pi_X(\cdot)$ is the projection mapping onto X; see also Figure 13.1. It follows now from Proposition

11.12 that iteration (13.1) is well-defined, moreover, x^{k+1} is also the unique solution of the following auxiliary VI: Find $x^{k+1} \in X$ such that

$$(x - x^{k+1})^T [G(x^k) + \lambda_k^{-1}(x^{k+1} - x^k)] \geq 0 \quad \forall x \in X. \tag{13.2}$$

Exercise 13.1. By using Proposition 11.12 (ii), prove the equivalence between (13.1) and (13.2).

To obtain convergence results for the projection method, we need certain additional assumptions, together with those in Definition 11.1.

Definition 13.1. A mapping $Q : X \to E$ is said to be

(a) *Lipschitz continuous* with constant L, if for each pair of points $x, y \in X$, we have

$$\|Q(x) - Q(y)\| \leq L\|x - y\|;$$

(b) *co-coercive* (or *inverse strongly monotone*) with constant $\mu > 0$, if for each pair of points $x, y \in X$, we have

$$(x - y)^T [Q(x) - Q(y)] \geq \mu \|Q(x) - Q(y)\|^2.$$

One of the strongest convergence results for method (13.1) can be formulated as follows.

Theorem 13.1. *Suppose that $G : X \to E$ is strongly monotone with constant $\tau > 0$ and Lipschitz continuous with constant L. If a sequence $\{x^k\}$ is constructed by the projection method (13.1) with $\lambda_k = \lambda \in (0, 2\tau/L^2)$, then it converges geometrically to the unique solution x^* of VI (11.1), i.e.*

$$\|x^{k+1} - x^*\| \leq \nu \|x^k - x^*\|, \quad \nu \in (0, 1). \tag{13.3}$$

Proof. First we note that VI (11.1) has now a unique solution because of Theorem 11.4. Using Propositions 11.12 and 11.13, we have

$$\begin{aligned}
&\|x^{k+1} - x^*\|^2 \\
&= \|\pi_X[x^k - \lambda_k G(x^k)] - \pi_X[x^* - \lambda_k G(x^*)]\|^2 \\
&\leq \|(x^k - x^*) - \lambda_k [G(x^k) - G(x^*)]\|^2 \\
&= \|x^k - x^*\|^2 - 2\lambda_k (x^k - x^*)^T [G(x^k) - G(x^*)] \\
&\quad + \lambda_k^2 \|G(x^k) - G(x^*)\|^2.
\end{aligned} \tag{13.4}$$

Now, using the assumptions of the theorem, we obtain

$$\begin{aligned}
\|x^{k+1} - x^*\|^2 &\leq \|x^k - x^*\|^2 - 2\lambda\tau \|x^k - x^*\|^2 \\
&\quad + \lambda^2 L^2 \|x^k - x^*\|^2 \\
&= (1 - \lambda(2\tau - \lambda L^2)) \|x^k - x^*\|^2 \\
&= \nu^2 \|x^k - x^*\|^2.
\end{aligned}$$

From the definition of λ it follows that $\nu = \sqrt{1 - \lambda(2\tau - \lambda L^2)} \in (0, 1)$, hence (13.3) is fulfilled. □

However, the conditions of Theorem 13.1 may appear too strong for applications (see Chapters 10–12) and we intend to prove similar results under weaker assumptions. In fact, we can replace the above conditions by co-coercivity of G. Observe that co-coercivity of G is equivalent to strong monotonicity of the inverse mapping G^{-1} and implies the Lipschitz continuity of G with constant $1/\mu$. Moreover, the conditions of Theorem 13.1 imply that G is co-coercive with constant $\mu = \tau/L^2$. At the same time, a co-coercive mapping need not be even strictly monotone in general.

Exercise 13.2. Let $G(x) = Ax + b$, where A is a symmetric positive semidefinite $n \times n$ matrix, $b \in \mathbb{R}^n$. Prove that G is co-coercive.

Let us consider the auxiliary mapping

$$T(x) = x - \pi_X \left[x - \lambda G(x)\right]. \tag{13.5}$$

Due to Proposition 11.13, its zeros coincide with X^*.

Lemma 13.1. *If G is co-coercive with constant $\mu > 0$, then T in* (13.5) *is co-coercive with constant $\mu' = 1 - \frac{\lambda}{4\mu}$ where $\lambda \in (0, 4\mu)$.*

Proof. Choose arbitrary points $x', x'' \in X$ and set $t' = T(x')$, $t'' = T(x'')$. Then, using Proposition 11.12, we have

$$\begin{aligned}
&[(x' - t') - (x'' - t'')]^T ([t' - \lambda G(x')] - [t'' - \lambda G(x'')]) \\
&= (\pi_X[x' - \lambda G(x')] - \pi_X[x'' - \lambda G(x'')])^T \\
&\times ([x' - \lambda G(x')] - [x'' - \lambda G(x'')]) \\
&- \|\pi_X[x' - \lambda G(x')] - \pi_X[x' - \lambda G(x'')]\|^2 \geq 0.
\end{aligned}$$

It follows that

$$\begin{aligned}
&(t' - t'')^T(x' - x'') \\
&\geq \|t' - t''\|^2 + \lambda(x' - x'')^T[G(x') - G(x'')] \\
&-\lambda(t' - t'')^T[G(x') - G(x'')] \\
&\geq \|t' - t''\|^2 + \lambda\mu\|G(x') - G(x'')\|^2 \\
&-\lambda(t' - t'')^T[G(x') - G(x'')] \\
&= (1 - \frac{\lambda}{4\mu})\|t' - t''\|^2 \\
&+ \left\|\sqrt{\frac{\lambda}{4\mu}}(t' - t'') - \sqrt{\lambda\mu}[G(x') - G(x'')]\right\|^2 \\
&\geq \mu'\|t' - t''\|^2,
\end{aligned}$$

as desired. □

This property ensures the convergence result.

Theorem 13.2. *Suppose that $G : X \to E$ is co-coercive with constant $\mu > 0$ and that VI (11.1) is solvable. If a sequence $\{x^k\}$ is constructed by the projection method (13.1) with $\lambda_k = \lambda \in (0, 2\mu)$, it converges to a point of X^*.*

Proof. Following (13.4) and using Lemma 13.1, we have

$$\begin{aligned}
&\|x^{k+1} - x^*\|^2 = \left\|[x^k - T(x^k)] - [x^* - T(x^*)]\right\|^2, \\
&= \|x^k - x^*\|^2 - 2(x^k - x^*)^T [T(x^k) - T(x^*)] \\
&+\|T(x^k) - T(x^*)\|^2 \\
&\leq \|x^k - x^*\|^2 - (2\mu' - 1)\|T(x^k) - T(x^*)\|^2 \\
&= \|x^k - x^*\|^2 - (2\mu' - 1)\|T(x^k)\|^2 \\
&\leq \|x^k - x^*\|^2
\end{aligned}$$

for an arbitrary solution x^* of VI (11.1) since

$$2\mu' - 1 = 1 + (1 - \frac{\lambda}{2\mu}) - 1 > 0.$$

It follows that the sequence $\{x^k\}$ is bounded, hence it has limit points, moreover,

$$\lim_{k\to\infty} T(x^k) = 0.$$

Taking an arbitrary limit point $\tilde{x}$ of $\{x^k\}$, we now obtain $T(\tilde{x}) = 0$ by continuity of G and the projection mapping. Due to Proposition 11.13, it means that $\tilde{x} \in X^*$, i.e. we can replace x^* by $\tilde{x}$ in the above inequalities and the monotone decrease of the distance $\|x^k - \tilde{x}\|$ yields

$$\lim_{k\to\infty} x^k = \tilde{x}.$$

□

We now present another way of weakening the conditions of Theorem 13.1. Let us consider the *acute angle condition*:

$$\forall x \in X \setminus X^*, \forall x^* \in X^*, \quad (x - x^*)^T G(x) > 0. \tag{13.6}$$

In fact, it means that the angle between the vectors $-G(x)$ and $x^* - x$ is acute at each non-optimal point x, i.e. the ray $\{z \mid z = x - \lambda G(x)\}$ leads to decrease of the distance to any optimal point x^*; see Figure 13.1. A proper choice of the stepsize then yields convergence of method (13.1), as the following theorem states.

Figure 13.1:

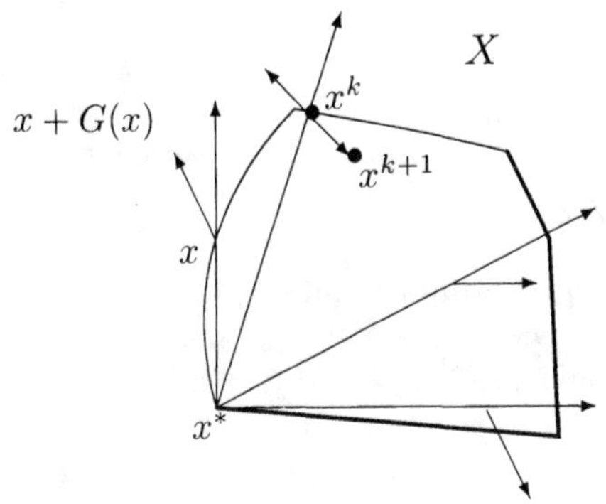

Theorem 13.3. *Suppose that VI (11.1) is solvable and that (13.6) holds. If a sequence $\{x^k\}$ is constructed by the projection method (13.1), where*

$$\lambda_k = \frac{\alpha_k}{\|G(x^k)\|}, \sum_{k=0}^{\infty} \alpha_k = \infty, \sum_{k=0}^{\infty} \alpha_k^2 < \infty, \tag{13.7}$$

then it either converges to a point of X^* or terminates at a point of X^*.

Proof. Clearly, if $G(x^k) = 0$, then $x^k \in X^*$. Consider the case when $G(x^k) \neq 0$ for $k = 0, 1, \ldots$ Fix $x^* \in X^*$, then, by construction,

$$\begin{aligned} \|x^{k+1} - x^*\|^2 &\leq \|\pi_X[x^k - \lambda_k G(x^k)] - \pi_X[x^*]\| \\ &\leq \|x^k - x^*\|^2 - 2\lambda_k (x^k - x^*)^T G(x^k) \\ &\quad + \lambda_k^2 \|G(x^k)\|^2 \\ &\leq \|x^k - x^*\|^2 + \alpha_k^2. \end{aligned}$$

Due to (13.7), we see that the sequence $\{x^k\}$ is bounded, hence it has limit points, moreover,

$$\lim_{k \to \infty} \|x^k - x^*\| = d \geq 0. \tag{13.8}$$

In fact, the numerical sequence $\{\delta_k\}$, $\delta_k = \|x^k - x^*\|^2$, also has limit points due to its boundedness. If δ' and δ'' are two different limit points of $\{\delta_k\}$

and $\delta' < \delta''$, then we can choose n such that $\sum_{k=n}^{\infty} \alpha_k^2 < (\delta'' - \delta')/4$ and $N \geq n$ such that $|\delta_N - \delta'| < (\delta'' - \delta')/4$. Then, by (13.7), for each $k > N$ we have

$$\begin{aligned}\delta_k &\leq \delta_N + \sum_{i=N}^{k-1} \alpha_i^2 \leq \delta' + |\delta_N - \delta'| + \sum_{i=n}^{k-1} \alpha_i^2 < \delta' + 0.5(\delta'' - \delta') \\ &= \delta'' - 0.5(\delta'' - \delta'),\end{aligned}$$

which is a contradiction since δ'' is a limit point of $\{\delta_k\}$. Therefore, $\delta' = \delta''$ and (13.8) holds.

Suppose that $d > 0$ and set $\mu_k = (x^k - x^*)^T G(x^k)/\|G(x^k)\| > 0$. If

$$\mu_k \geq \mu' > 0 \quad \text{for} \quad k = 0, 1, \ldots,$$

then, by the above inequalities,

$$\begin{aligned}\|x^{k+1} - x^*\|^2 &\leq \|x^k - x^*\|^2 - 2\alpha_k \mu' + \alpha_k^2 \\ &\leq \|x^0 - x^*\|^2 - 2\mu' \sum_{i=0}^{k} \alpha_i + \sum_{i=0}^{k} \alpha_i^2 < 0\end{aligned}$$

for sufficiently large k, a contradiction. Therefore, there exists a sequence $\{k_s\}$ such that $\mu_{k_s} \to 0$ as $k_s \to +\infty$. Then, taking a subsequence, if necessary, we obtain

$$\begin{aligned}0 = \lim_{k_s \to \infty} \mu_{k_s} &= \lim_{k_s \to \infty} \left[(x^{k_s} - x^*)^T G(x^{k_s})/\|G(x^{k_s})\|\right] \\ &= (x' - x^*)^T G(x')/\|G(x')\|,\end{aligned}$$

where x' is the corresponding limit point of $\{x^{k_s}\}$. It follows now from (13.6) that $x' \in X^*$. Since $x^* \in X^*$ was taken arbitrarily, we conclude that $d = 0$ in (13.8), i.e.

$$\lim_{k \to \infty} \|x^k - x'\| = 0,$$

and the result follows. □

Observe that condition (13.6) can be viewed as an extension of the revealed preference condition (10.6), and the process (13.1), (13.7) is the corresponding extension of the Walras tâtonnement (cf. (10.4), (10.5)). It is easy to see that strict monotonicity of G implies (13.6), however, co-coercivity and the acute angle conditions do not imply each other. A more detailed description of properties of the classical projection method and its extensions under these conditions can be found in the book by *Gol'shtein and Tret'yakov* (1989).

Figure 13.2:

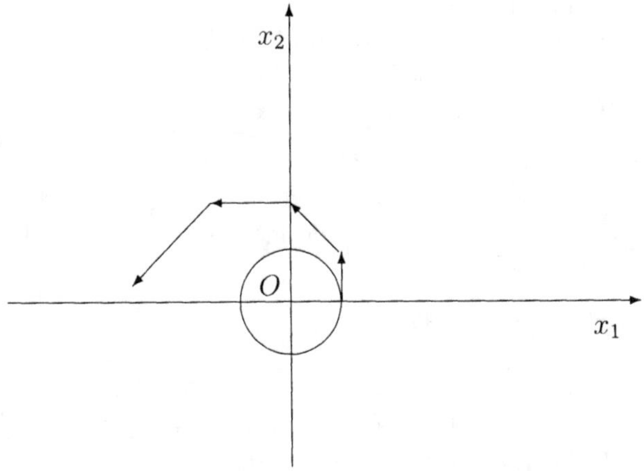

It should be also observed that the projection method (13.1) does not provide convergence in the case when G is a general monotone mapping; see e.g. *Konnov* (2001), Example 1.2.1.

Example 13.1. Set $E = X = \mathbb{R}^2$, $G(x) = (x_2, -x_1)^T$. Then G is clearly monotone and $X^* = \{(0,0)^T\}$. For any $\lambda_k > 0$, we have

$$\begin{aligned} \|x^k - \lambda_k G(x^k)\|^2 &= (x_1^k - \lambda_k x_2^k)^2 + (x_2^k + \lambda_k x_1^k)^2 \\ &= (1 + \lambda_k^2)\|x^k\|^2 > \|x^k\|^2; \end{aligned}$$

i.e., the distance to the solution may only increase regardless of the stepsize rule; see Figure 13.2.

This property is caused by the fact that the angle between $-G(x^k)$ and $x^* - x^k$ need not be acute in the general monotone case (cf. (13.6)).

Exercise 13.3. Prove that assumption (13.6) is not fulfilled if $G(x) = Ax + b$, where A is an arbitrary skew-symmetric matrix.

At the same time, many equilibrium problems are formulated as VIs without strengthened monotonicity assumptions. Some methods ensuring convergence for such VIs will be described in Chapters 15 and 16.

13.2 The projection methods with linesearch

We now describe a way of constructing projection-based iterations (13.1) with incorporation of a suitable linesearch procedure. This approach does not require a priori knowledge of constants of G, such as Lipschitz continuity, co-coercivity and strong monotonicity ones. However, such a linesearch procedure requires a merit function. For this reason, we first consider the case when G is integrable, i.e.

$$G(x) = \nabla f(x) \tag{13.9}$$

for a function $f : \mathbb{R}^n \to \mathbb{R}$, see Section 11.1. Let us consider the auxiliary mapping

$$Z(x) = \pi_X[x - \lambda \nabla f(x)] \tag{13.10}$$

for some $\lambda > 0$.

The next lemma collects the basic properties of Z, which ensure convergence of the gradient projection method.

Lemma 13.2. *Let Z be defined by (13.10) and let (13.9) hold. Then:*
(i) $x \mapsto Z(x)$ is continuous;
(ii) $x^* \in X^* \iff x^* = Z(x^*)$;
(iii) $x \in X \Longrightarrow (Z(x) - x)^T \nabla f(x) \leq -\lambda \|Z(x) - x\|^2$.

Proof. The continuity of Z follows from the continuity of G and $\pi_X(\cdot)$. Proposition 11.12 shows that (13.10) is equivalent to

$$\begin{array}{l} Z(x) \in X, \\ [y - Z(x)]^T \left(\nabla f(x) + \lambda^{-1}[Z(x) - x]\right) \geq 0 \quad \forall y \in X \end{array} \tag{13.11}$$

(cf. (13.1) and (13.2)). Therefore, $x^* = Z(x^*)$ gives $x^* \in X^*$ by definition. Next, setting $y = x$ in (13.11) yields

$$-\lambda^{-1}\|x - Z(x)\|^2 \geq (Z(x) - x)^T \nabla f(x),$$

i.e. (iii) holds true, and $x \neq Z(x)$ implies $x \notin X^*$. Hence, (ii) is also true. □

We now describe the descent gradient projection algorithm for VI (11.1), (13.9).

Algorithm (descent gradient projection). Choose a point $x^0 \in X$, numbers $\beta \in (0, 1)$, $\gamma \in (0, 1)$, $\lambda > 0$, and set $k = 0$.

At the k-th iteration, we have a point $x^k \in X$, compute $z^k = Z(x^k)$ and set $d^k = z^k - x^k$. If $d^k = 0$, stop. Otherwise, find m as the smallest non-negative integer such that

$$f(x^k + \gamma^m d^k) \leq f(x^k) + \beta\gamma^m (d^k)^T \nabla f(x^k), \tag{13.12}$$

set $\mu_k = \gamma^m, x^{k+1} = x^k + \mu_k d^k$, and $k = k+1$.

According to the description, the method finds a solution to VI in the case of its finite termination, i.e. $d^k = 0$ implies $x^k \in X^*$ on account of Lemma 13.2 (ii). For this reason, in what follows we shall consider only the case of the infinite sequence $\{x^k\}$. The convergence of this method can be established under rather weak additional assumptions.

Theorem 13.4. *Suppose that (13.9) holds and that the set* $L_0 = \{x \in X \mid f(x) \leq f(x^0)\}$ *is bounded. Then the sequence* $\{x^k\}$ *generated by the above algorithm is well-defined and have limit points, moreover, all these limit points are solutions of VI (11.1), (13.9).*

Proof. First we note that f is continuous due to (13.9), hence, by the *Weierstrass* theorem, there exists a solution of the optimization problem (11.5), and Theorem 11.1 now yields solvability of VI (11.1), (13.9). Next, since f is differentiable, applying the *Taylor* formula gives

$$f(x + \mu d^k) - f(x^k) = \mu (d^k)^T \nabla f(x^k) + o(\mu)$$

for $\mu \geq 0$ and Lemma 13.2 (iii) yields $(d^k)^T \nabla f(x^k) < 0$. Hence, there exists $\mu' > 0$ such that

$$f(x^k + \mu d^k) - f(x^k) \leq \beta\mu (d^k)^T \nabla f(x^k)$$

for all $\mu \in (0, \mu')$ and the linesearch procedure with condition (13.12) is well-defined. It follows that the sequence $\{f(x^k)\}$ is decreasing and that the sequence $\{x^k\}$ is contained in the bounded set L_0, hence $\{x^k\}$ has limit points. Suppose that $\lim\limits_{k\to\infty} d^k \neq 0$, then there exists a subsequence $\{\mu_{k_s}\}$ such that $\{\mu_{k_s}\} \to 0$ as $\{k_s\} \to +\infty$. Then, for $k = k_s$ large enough, we have

$$f(x^k + (\mu_k/\gamma) d^k) - f(x^k) > \beta (\mu_k/\gamma) (d^k)^T \nabla f(x^k)$$

or equivalently,

$$(\mu_k/\gamma) \left[f(x^k + (\mu_k/\gamma) d^k) - f(x^k) \right] > \beta (d^k)^T \nabla f(x^k).$$

Taking the limit $k = k_s \to +\infty$ and a subsequence, if necessary, we obtain

$$\tilde{d}^T \nabla f(\tilde{x}) \geq \beta \tilde{d}^T \nabla f(\tilde{x}),$$

where $\tilde{x}$ and $\tilde{d}$ are the corresponding limit points of $\{x^k\}$ and $\{d^k\}$, hence $\tilde{d}^T \nabla f(\tilde{x}) \geq 0$. But $\tilde{d} \neq 0$ and $\tilde{x}$ does not solve VI (11.1), (13.9), i.e. $\tilde{d}^T \nabla f(\tilde{x}) < 0$ because of Lemma 13.2, a contradiction. Therefore,

$$\lim_{k \to \infty} (Z(x^k) - x^k) = 0,$$

and Lemma 13.2 now gives that all the limit points of $\{x^k\}$ belong to X^*. □

Theorem 13.4 shows that the integrability of the mapping G enables us to construct a converging iteration sequence without any monotonicity assumptions. However, the integrability seems too strong for many VIs arising in economic equilibrium problems. In order to construct such a descent method in the general case we need an artificial merit function and certain strengthened monotonicity conditions for its convergence.

Let us consider VI (11.1) where $G : X \to \mathbb{R}^n$ is a continuously differentiable mapping. Observe that the so-called primal merit function

$$\varphi(x) = \sup_{y \in X} (x - y)^T G(x)$$

provides the equivalence between VI (11.1) and the optimization problem

$$\min_{x \in X} \to \varphi(x),$$

however, it is not so easy to find a suitable algorithm for the latter problem due to the absence of convexity and differentiability of φ. The regularized merit function suggested by *M. Fukushima* possesses better differentiability properties. It is defined as follows:

$$\varphi_\lambda(x) = \max_{y \in X} \left\{ (x - y)^T G(x) - (2\lambda)^{-1} \|x - y\|^2 \right\} \tag{13.13}$$

for a fixed number $\lambda > 0$. Since the inner problem in (13.13) is a strongly concave maximization one, it has a unique solution, which is denoted by $Z_\lambda(x)$. In fact, writing the necessary and sufficient condition of optimality for this problem from Theorem 11.1, we see that $Z_\lambda(x) \in X$ and

$$(y - Z_\lambda(x))^T (G(x) + \lambda^{-1}[Z_\lambda(x) - x]) \geq 0 \quad \forall y \in X, \tag{13.14}$$

but Proposition 11.12 now gives

$$Z_\lambda(x) = \pi_X[x - \lambda G(x)] \tag{13.15}$$

(cf. (13.10)). Therefore, solving the inner problem in (13.13) is equivalent to the projection iteration (13.15), hence the mapping $x \mapsto Z_\lambda(x)$ is continuous.

We now show that φ_λ can serve as merit function for VI (11.1).

Proposition 13.1. *(i)* $\varphi_\lambda(x) \geq 0$ *for all* $x \in X$.
(ii) The following statements are equivalent:
(a) $\varphi_\lambda(x^*) = 0$ *and* $x^* \in X$,
(b) x^* *solves VI (11.1),*
(c) $x^* = Z_\lambda(x^*)$.

Proof. By definition,

$$\varphi_\lambda(x) \geq (x - x)^T G(x) - (2\lambda)^{-1}\|x - x\|^2 = 0 \quad \forall x \in X,$$

i.e. (i) is true. Next, the equivalence of (b) and (c) has been proven in Proposition 11.13. Moreover, (c) clearly implies (a). Conversely, let (a) hold true. Then, using (13.14) with $x = y = x^*$, we have

$$\begin{aligned} 0 &= \varphi_\lambda(x^*) \\ &= (x^* - Z_\lambda(x^*))^T G(x^*) - (2\lambda)^{-1}\|x^* - Z_\lambda(x^*)\|^2 \\ &\geq (2\lambda)^{-1}\|x^* - Z_\lambda(x^*)\|^2 \geq 0, \end{aligned}$$

i.e. $x^* = Z_\lambda(x^*)$ and (c) holds. The proof is complete. □

Proposition 13.1 enables us to replace VI (11.1) by the optimization problem

$$\min_{x \in X} \to \varphi_\lambda(x). \tag{13.16}$$

However, Theorem 13.4 says that the descent methods ensure convergence only to solutions of the associated VI: Find $x^* \in X$ such that

$$(y - x^*)^T \nabla\varphi_\lambda(x^*) \geq 0 \quad \forall y \in X. \tag{13.17}$$

Under the above assumptions, φ_λ is a differentiable function, and using the formula for the derivative of max type functions (see e.g. *Facchinei and Pang* (2003), Theorem 10.2.1), we have

$$\nabla\varphi_\lambda(x) = G(x) + [\nabla G(x) - \lambda^{-1} I]^T (x - Z_\lambda(x)). \tag{13.18}$$

We are now ready to obtain the equivalence result.

Proposition 13.2. *If* G *is differentiable and* $\nabla G(x)$ *is positive definite for each* $x \in X$, *then problems (11.1), (13.16), and (13.17) are equivalent.*

Proof. The equivalence of (11.1) and (13.16) has been proven in Proposition 13.1, moreover, Theorem 11.1 yields (13.16) $\Longrightarrow$ (13.17). Suppose that x^* solves (13.17), then

$$\begin{aligned} &(Z_\lambda(x^*) - x^*)^T \nabla\varphi_\lambda(x^*) \\ &= (Z_\lambda(x^*) - x^*)^T G(x^*) \\ &+ (Z_\lambda(x^*) - x^*)^T \nabla G(x^*)^T (x^* - Z_\lambda(x^*)) \\ &+ \lambda^{-1}\|x^* - Z_\lambda(x^*)\|^2 \geq 0. \end{aligned} \tag{13.19}$$

Applying (13.14) with $y = x^*, x = x^*$, we obtain

$$(Z_\lambda(x^*) - x^*)^T \nabla G(x^*)^T (x^* - Z_\lambda(x^*)) \geq 0,$$

hence $x^* = Z_\lambda(x^*)$ and x^* solves (11.1) because of Proposition 13.1. □

Therefore, applying the usual gradient projection method with respect to problem (13.17), i.e. replacing f by φ_λ, we can obtain an iterative sequence converging to a solution of the initial VI (11.1) if $\nabla G(x)$ is positive definite and the set $L_0 = \{x \in X \mid \varphi_\lambda(x) \leq \varphi_\lambda(x_0)\}$ is bounded. In fact, the positive definiteness of $\nabla G(x)$ implies the strict monotonicity of G on account of Proposition 11.1 (ii), i.e. the uniqueness of the solution of VI (11.1) because of Proposition 11.14. Following Theorem 13.4, we conclude that the iterative sequence converges to this unique solution. However, (13.19) and (13.14) then imply

$$\begin{aligned}
&(Z_\lambda(x) - x)^T \nabla \varphi_\lambda(x) = (Z_\lambda(x) - x)^T G(x) \\
&+ (Z_\lambda(x) - x)^T \nabla G(x)^T (x - Z_\lambda(x)) + \lambda^{-1} \|x - Z_\lambda(x)\|^2 \\
&< (Z_\lambda(x) - x)^T \left[G(x) + \lambda^{-1}(Z_\lambda(x) - x) \right] < 0
\end{aligned}$$

if $\nabla G(x)$ is positive definite and $Z_\lambda(x) \neq x$, hence $Z_\lambda(x) - x$ is also a descent direction for φ_λ at each non-optimal point x. Being based on this observation, we can propose a descent projection algorithm.

Algorithm (descent projection). Choose a point $x^0 \in X$, numbers $\beta \in (0, 1)$, $\gamma \in (0, 1)$, $\lambda > 0$, and set $k = 0$.

At the k-th iteration, we have a point $x^k \in X$, compute $z^k = Z_\lambda(x^k)$ and set $d^k = z^k - x^k$. If $d^k = 0$, stop. Otherwise, find m as the smallest non-negative integer such that

$$\varphi_\lambda(x^k + \gamma^m d^k) \leq \varphi_\lambda(x^k) + \beta \gamma^m (d^k)^T \nabla \varphi_\lambda(x^k), \tag{13.20}$$

set $\mu_k = \gamma^m, x^{k+1} = x^k + \mu_k d^k$, and $k = k + 1$.

Since $d^k = 0$ implies $x^k \in X^*$, we shall consider only the case of the infinite sequence $\{x^k\}$. Taking into account Theorem 13.4 and the above observations, we can obtain the following convergence property of this algorithm.

Theorem 13.5. *Suppose that the set $L_0 = \{x \in X \mid \varphi_\lambda(x) \leq \varphi_\lambda(x^0)\}$ is bounded and that $\nabla G(x)$ is positive definite at each point x of X. Then the sequence $\{x^k\}$ generated by the above algorithm is well-defined and converges to a unique solution of VI (11.1).*

Exercise 13.4. Prove the assertion of Theorem 13.5.

We see that the computation of the descent direction in the above algorithm does not require the computation of the Jacobian $\nabla G(x)$, however this is not the case for the verification of rule (13.20); see (13.18). Under the strong monotonicity assumption, we can construct the modified descent algorithm without computation of the Jacobian.

Algorithm (modified descent projection). Choose a point $x^0 \in X$, numbers $\beta \in (0,1)$, $\gamma \in (0,1)$, $\lambda > 0$, and set $k = 0$.

At the k-th iteration, we have a point $x^k \in X$, compute $z^k = Z_\lambda(x^k)$ and set $d^k = z^k - x^k$. If $d^k = 0$, stop. Otherwise, find m as the smallest non-negative integer such that

$$\varphi_\lambda(x^k + \gamma^m d^k) \leq \varphi_\lambda(x^k) - \beta\gamma^m \|d^k\|^2, \tag{13.21}$$

set $\mu_k = \gamma^m$, $x^{k+1} = x^k + \mu_k d^k$, and $k = k + 1$.

First we establish the basic descent property similar to that in Lemma 13.2 (iii).

Lemma 13.3. *If G is strongly monotone with constant τ, then*

$$(Z_\lambda(x) - x)^T \nabla\varphi_\lambda(x) \leq -\tau\|Z_\lambda(x) - x\|^2 \quad \forall x \in X. \tag{13.22}$$

Proof. By definition,

$$\begin{aligned}
&(Z_\lambda(x) - x)^T \nabla\varphi_\lambda(x) = (Z_\lambda(x) - x)^T G(x)\\
&+(Z_\lambda(x) - x)^T \nabla G(x)^T (x - Z_\lambda(x))\\
&+\lambda^{-1}\|x - Z_\lambda(x)\|^2\\
&\leq (Z_\lambda(x) - x)^T \nabla G(x)^T (x - Z_\lambda(x))\\
&\leq -\tau\|Z_\lambda(x) - x\|^2,
\end{aligned}$$

where the first inequality follows from (13.4) with $y = x$, and the second inequality follows from Proposition 11.1 (iii). Therefore, (13.22) is true. □

We additionally establish the error bound using the merit function φ_λ.

Lemma 13.4. *If G is strongly monotone with constant τ, then there exists a number $\sigma > 0$ such that*

$$\varphi_\lambda(x) \geq \sigma\|x - x^*\|^2 \quad \forall x \in X. \tag{13.23}$$

where x^* is the unique solution of VI (11.1).

Proof. First we note that VI (11.1) has now the unique solution x^* due to Theorem 11.4. Choose $\alpha \in (0,1)$ and an arbitrary point $x \in X$ and set $y_\alpha = \alpha x^* + (1-\alpha)x$. By definition, we have

$$\begin{aligned}\varphi_\lambda(x) &\geq (x-y_\alpha)^T G(x) - (2\lambda)^{-1}\|x-y_\alpha\|^2 \\ &= \alpha(x-x^*)^T G(x) - \alpha^2(2\lambda)^{-1}\|x-x^*\|^2,\end{aligned}$$

The strong monotonicity of G gives

$$\begin{aligned}(x-x^*)^T G(x) &\geq (x-x^*)^T G(x^*) + \tau\|x-x^*\|^2 \\ &\geq \tau\|x-x^*\|^2,\end{aligned}$$

hence

$$\varphi_\lambda(x) \geq \alpha\left(\tau - \alpha(2\lambda)^{-1}\right)\|x-x^*\|^2$$

for arbitrary $\alpha \in (0,1)$, i.e. (13.23) holds with

$$\sigma = \begin{cases} \tau - (2\lambda)^{-1} & \text{if } \tau \geq 1/\lambda, \\ \tau^2\lambda/2 & \text{if } \tau < 1/\lambda. \end{cases}$$

□

These results provide convergence of the above algorithm. Again, we shall consider only the case of the infinite sequence $\{x^k\}$.

Theorem 13.6. *Suppose that the mapping G is continuously differentiable and strongly monotone with constant $\tau > 0$. Then the sequence $\{x^k\}$ generated by the above algorithm with $\beta < \tau$ is well-defined and converges to a unique solution of VI (11.1).*

Proof. Again, Theorem 11.4 yields the existence and uniqueness of the solution of VI (11.1). Since φ_λ is now differentiable, applying the Taylor formula gives

$$\varphi_\lambda(x^k + \mu d^k) - \varphi_\lambda(x^k) = \mu(d^k)^T \nabla\varphi_\lambda(x^k) + o(\mu)$$

for $\mu \geq 0$ and (13.22) yields $(d^k)^T \nabla\varphi_\lambda(x^k) < -\tau\|d^k\|^2$. Hence, there exists $\mu' > 0$ such that

$$\varphi_\lambda(x^k + \mu d^k) - \varphi_\lambda(x^k) \leq -\beta\mu\|d^k\|^2$$

for all $\mu \in (0, \mu')$ and the linesearch procedure with condition (13.21) is always finite, i.e. the algorithm is well-defined. Rule (13.21) implies that the sequence $\{\varphi_\lambda(x^k)\}$ is decreasing, but the set $L_0 = \{x \in X \mid \varphi_\lambda(x) \leq \varphi(x^0)\}$ is bounded due to (13.23). Hence, the sequence $\{x^k\}$ has limit points. Suppose that $\lim_{k\to\infty} d^k \neq 0$, then there exists a subsequence $\{\mu_{k_s}\} \to 0$ as $\{k_s\} \to +\infty$. Then, for $k = k_s$ large enough, we have

$$\begin{aligned}\varphi_\lambda(x^k + (\mu_k/\gamma)d^k) &- \varphi_\lambda(x^k) > -\beta(\mu_k/\gamma)\|d^k\|^2 \\ &\geq \beta(\mu_k/\gamma)\tau^{-1}(d^k)^T\nabla\varphi_\lambda(x^k)\end{aligned}$$

due to (13.22), hence

$$(\mu_k/\gamma)\left[\varphi_\lambda(x^k + (\mu_k/\gamma)d^k) - \varphi_\lambda(x^k)\right] > (\beta/\tau)(d^k)^T\nabla\varphi_\lambda(x^k).$$

Taking the limit $k = k_s \to +\infty$ and a subsequence, if necessary, we obtain

$$(\tilde{d})^T\nabla\varphi_\lambda(\tilde{x}) \geq (\beta/\tau)(\tilde{d})^T\nabla\varphi_\lambda(\tilde{x}),$$

where $\tilde{x}$ and $\tilde{d}$ are the corresponding limit points of $\{x^k\}$ and $\{d^k\}$, hence

$$(\tilde{d})^T\nabla\varphi_\lambda(\tilde{x}) \geq 0.$$

But $\tilde{d} \neq 0$ and (13.22) yields

$$(\tilde{d})^T\nabla\varphi_\lambda(\tilde{x}) \leq -\tau\|\tilde{d}\|^2 < 0,$$

a contradiction. Therefore, $\lim_{k\to\infty} d^k = 0$ and, on account of the continuity of Z_λ, each limit point of $\{x^k\}$ is a solution of VI (11.1). However, the solution point is unique and $\{x^k\}$ must converge to this point. □

13.3 Modifications and extensions

The projection method described in the previous sections admits modifications and extensions in several directions. One of them consists in replacing the simplest affine approximation of the mapping G at the current point x^k, given in (13.2), by more general expressions. In fact, we can consider the method where the next iteration point is a solution of the following auxiliary VI: Find $x^{k+1} \in X$ such that

$$(x - x^{k+1})^T[G(x^k) + \lambda_k^{-1}A_k(x^{k+1} - x^k)] \geq 0 \quad \forall x \in X, \tag{13.24}$$

where A_k is an $n \times n$ matrix, so that the projection method corresponds to the choice $A_k \equiv I$. If A_k is positive definite, then the mapping $\tilde{G}^k(x) = G(x^k) + \lambda_k^{-1}A_k(x - x^k)$ is strongly monotone and VI (13.24) has a unique solution; see Theorem 11.4. Then the method becomes well-defined. For instance, if G is strongly monotone, then, by Proposition 11.1 (iii), its Jacobian $\nabla G(x)$ is positive definite and we can set $A_k = \nabla G(x^k)$ and obtain a version of the well-known Newton method, which possesses very fast local convergence. Therefore, taking A_k as a suitable approximation of $\nabla G(x^k)$, we can obtain a number of so-called quasi-Newton methods, whose properties are intermediate between the projection and Newton methods. Their convergence results are based partially on the techniques utilized in the substantiation of the projection method, but construction of more efficient versions requires a more complicated and deeper analysis. The detailed description of such methods is presented in the books by *Ortega and Rheinboldt* (1970), *Patriksson* (1999), and *Facchinei and Pang* (2003).

Another approach to extending the projection iteration (13.1) consists in replacing the projection mapping $\pi_X(\cdot)$ with a somewhat more general mapping possessing the properties indicated in Propositions 11.12 and 11.13. Then the corresponding extension of the projection method will have similar convergence properties.

Let $Q : X \to \mathbb{R}^n$ be a monotone continuous mapping and let $\lambda > 0$ be a fixed number. Then, for each point $x \in \mathbb{R}^n$, there exists a unique solution of the following VI: Find $P(x) \in X$ such that

$$(y - P(x))^T \left(Q[P(x)] + \lambda^{-1}[P(x) - x]\right) \geq 0 \quad \forall y \in X, \tag{13.25}$$

thus determining the so-called *proximal mapping* $x \mapsto P(x)$ with respect to the mapping Q; this mapping was first introduced by *J.J. Moreau* for the case when Q is integrable. In fact, the mapping $Q(z) + \lambda^{-1}(z - x)$ is strongly monotone and continuous and the above assertion follows from Theorem 11.4.

We now obtain the basic properties of the proximal mapping.

Proposition 13.3. *Let $P : \mathbb{R}^n \to X$ be the proximal mapping with respect to a monotone continuous mapping Q and $\lambda > 0$. Then P is non-expansive and continuous, and*

$$\begin{array}{c}(x'' - x')^T[P(x') - P(x'')] \geq \|P(x') - P(x'')\|^2 \\ \forall x', x'' \in \mathbb{R}^n.\end{array} \tag{13.26}$$

Proof. Choose arbitrary points $x', x'' \in \mathbb{R}^n$ and set $p' = P(x')$, $p'' = P(x'')$. Then, by (13.25), we have

$$(p'' - p')^T[Q(p') + \lambda^{-1}(p' - x')] \geq 0$$

and

$$(p' - p'')^T[Q(p'') + \lambda^{-1}(p'' - x'')] \geq 0.$$

Adding these inequalities gives

$$\begin{aligned}&(p'' - p')^T(x'' - x') \\ &\geq \lambda(p'' - p')^T[Q(p'') - Q(p')] + \|p'' - p'\|^2 \\ &\geq \|p'' - p'\|^2,\end{aligned}$$

i.e. (13.26) holds true. Next, applying the Cauchy-Schwarz inequality in (13.26) yields

$$\|x'' - x'\| \geq \|P(x'') - P(x')\|,$$

i.e. P is non-expansive, which in turn implies the continuity of P. The proof is complete. □

Now, let us consider the initial VI (11.1) and suppose that

$$G(x) = F(x) + Q(x), \tag{13.27}$$

where $Q : X \to \mathbb{R}^n$ is a monotone continuous mapping and F is a continuous mapping. Then we can replace (13.1) by the following rule:

$$x^{k+1} = P[x^k - \lambda F(x^k)], \tag{13.28}$$

and, due to (13.25), (13.28) is then rewritten as follows: Find $x^{k+1} \in X$ such that

$$\begin{aligned}&(y - x^{k+1})^T \left[F(x^k) + \lambda^{-1}(x^{k+1} - x^k)\right.\\ &\left.+Q(x^{k+1})\right] \geq 0 \quad y \in X;\end{aligned} \tag{13.29}$$

cf. (13.2). Both rules (13.28) and (13.29) represent the same method, which is explicit with respect to F and implicit with respect to Q. It is called the *forward-backward splitting* method and is due to *P.L. Lions and B. Mercier.* Observe that setting $Q \equiv 0$ in (13.27) and (13.29) leads to the previous projection method since $G = F$. Thus, the splitting method is an extension of the projection method, but its implementation requires that the auxiliary problem (13.29) would be solved rather easily. The next property shows that the initial VI (11.1) is equivalent to a fixed point problem defined with the help of the proximal mapping.

Proposition 13.4. *If $Q : X \to \mathbb{R}^n$ is a monotone continuous mapping, then a point $x^* \in X$ is a solution of VI (11.1), (13.27) if and only if*

$$x^* = P[x^* - \lambda F(x^*)]. \tag{13.30}$$

Proof. If (13.30) holds, then (13.25) with $x = x^* - \lambda F(x^*)$ gives

$$(y - x^*)^T [Q(x^*) + F(x^*)] \geq 0 \quad \forall y \in X,$$

i.e. x^* is a solution of VI (11.1), (13.27). Conversely, the above inequality implies

$$(y - x^*)^T [Q(x^*) + \lambda^{-1} (x^* - (x^* - \lambda F(x^*)))] \geq 0 \quad y \in X,$$

i.e. $x^* = P[x^* - \lambda F(x^*)]$ because of (13.25). □

The above properties of the proximal mapping are similar to those of the projection mapping in Propositions 11.12 and 11.13. Therefore, we can obtain the convergence of the splitting method (13.28) (or (13.29)) with fixed stepsize along the lines of Theorems 13.1 and 13.2.

Theorem 13.7. *Suppose that (13.27) holds, where* $Q : X \to \mathbb{R}^n$ *is continuous and monotone and* $F : X \to \mathbb{R}^n$ *is Lipschitz continuous with constant* L *and strongly monotone with constant* τ. *If a sequence* $\{x^k\}$ *is constructed by the splitting method (13.28) with* $\lambda \in (0, 2\tau/L^2)$, *then it converges geometrically to a unique solution of VI (11.1), (13.27).*

Exercise 13.5. Prove the assertion of Theorem 13.7 along the lines of the proof of Theorem 13.1.

Similarly, we can obtain convergence of the above method under the co-coercivity of F.

Theorem 13.8. *Suppose that (13.27) holds, where* $Q : X \to \mathbb{R}^n$ *is continuous and monotone and* $F : X \to \mathbb{R}^n$ *is co-coercive with constant* $\mu > 0$, *and that VI (11.1), (13.27) is solvable. If a sequence* $\{x^k\}$ *is constructed by the splitting method (13.28) with* $\lambda \in (0, 2\mu)$, *then it converges to a solution of VI (11.1), (13.27).*

Exercise 13.6. Prove the assertion of Theorem 13.8 along the lines of the proofs of Theorem 13.2 and Lemma 13.1.

Observe that the splitting method enables us to weaken the conditions on the part Q of the cost mapping G. Moreover, the same convergence results can be obtained even for the case when Q is a multi-valued monotone mapping. The detailed description of various splitting methods and their convergence properties is given e.g. in the books by *Cottle, Pang, and Stone* (1992), *Patriksson* (1999), and *Facchinei and Pang* (2003).

Chapter 14

Applications of the Projection Methods

In this chapter, we discuss possible applications of the projection type methods to equilibrium problems described mostly in Part I. Some of them have been presented in Chapter 10. Now our analysis are based on the properties given in Chapters 11–13.

14.1 Applications to variational inequalities

We first consider the models that are formulated as the usual VI: Find $x^* \in X$ such that

$$(x - x^*)^T G(x^*) \geq 0 \quad \forall x \in X, \tag{14.1}$$

where X is a nonempty, convex and closed subset in $\mathbb{R}^n$, $G : X \to \mathbb{R}^n$ is a continuous mapping. The application of the projection method (13.1) tacitly assumes that the projection onto X can be implemented rather easily. Such equilibrium models were presented in Part I and analyzed in Chapter 12. Namely, the Cassel-Wald model (see (12.1)), the Walrasian type models (12.4) and (12.8), the oligopolistic model (11.1), (12.20), the path flow formulation of the network equilibrium model (12.22), (12.23), and the migration equilibrium model (6.9), (6.12) reduce to VI (14.1) with the desired properties of the feasible set X. Moreover, under rather natural assumptions, the corresponding cost mapping G then possesses the monotonicity properties; see Chapter 12. Therefore, we can apply the projection method (13.1) both for finding their solutions and for describing dynamic processes in there models.

For instance, in the Cassel-Wald model (12.1), x denotes the vector of outputs of commodities, and $G(x) = -c(x)$, where $c(x)$ denotes the price vector at x. Therefore, the (strengthened) monotonicity properties of $-c(x)$ seem very natural but they imply convergence of the projection method. Since the feasible set is defined by the linear constraints, i.e.,

$$X = \{x \in \mathbb{R}^n \mid Ax \leq b, x \geq 0\},$$

the projection onto X is implementable in the sense that it can be always computed in the finite number of operations.

If we consider the projection method (13.1) as a model of the dynamic process, it can be interpreted as a process with reaction only on the current state x^k and its estimate $G(x^k)$. For the Cassel-Wald model, it means that the change of the prices c only depends on the current output x^k without taking into account the history of the system. The results of Chapter 13 describe the conditions of stability of this process.

Exercise 14.1. Give interpretations of the projection method for the Walrasian type models (12.4) and (12.8), the oligopolistic model (11.1), (12.20), the auction market model (5.28), (5.29), the path flow formulation of the network equilibrium model (12.22), (12.23), and the migration equilibrium model (6.9), (6.12).

Most convergence theorems for the projection method require the strengthened monotonicity properties of G. In general, if G is only monotone, we should make use of some other methods described in the next chapters. At the same time, Theorem 13.4 says that the monotonicity may be in principle replaced by the integrability. For instance, this is the case if G is diagonal, i.e.

$$G(x) = (G_1(x_1), \dots, G_n(x_n))^T, \tag{14.2}$$

or G is an affine mapping with symmetric matrix, i.e.

$$\begin{array}{l} G(x) = Ax + b, \quad \text{where} \\ \ A \text{ is a } n \times n \text{ symmetric matrix.} \end{array} \tag{14.3}$$

Exercise 14.2. Prove that both (14.2) and (14.3) imply (13.9), i.e. the integrability of G.

For instance, if the price c_i of the i-th commodity in the Cassel-Wald model only depends on the output x_i of just this commodity, then $G(x) = -c(x)$ is diagonal, and the convergence of the projection method in this case follows from Theorem 13.4.

Exercise 14.3. Give the interpretation of the diagonal property of the cost mappings in the models indicated in Exercise 14.1.

Let us consider additionally the path flow formulation of the network equilibrium model (12.22), (12.23). It was noticed in Section 12.5 that (12.22) may be rewritten equivalently as VI (14.1), where

$$G(x) = B^T F(Bx), \tag{14.4}$$

B is the arc-path incidence matrix, x is the vector of path flows, $F(y)$ is the vector value of flow costs depending on the arc flows y; also, X in (12.23) is nonempty, convex, and compact; see (12.24). It is natural to suppose that the mapping F is (strongly, strictly) monotone, but this is not the case for G, however, G maintains both monotonicity and integrability of F. We refer to Exercise 12.7 for the monotone case.

Exercise 14.4. Prove that the integrability of F implies the integrability of G in (14.4).

Let now F be strongly monotone with constant τ and Lipschitz continuous with constant L. Since B is not a square matrix and contains zero entries, G in (14.4) need not be strongly monotone, but possesses co-coercivity with constant

$$\mu = \tau/(L\|B\|)^2.$$

In fact, fix $x', x'' \in X$ and set $y' = Bx'$, $y'' = Bx''$. Then we have

$$\begin{aligned}
&(x' - x'')^T[G(x') - G(x'')] \\
&= (y' - y'')^T[F(y') - F(y'')] \\
&\geq \tau\|y' - y''\|^2 \geq (\tau/L^2)\|F(y') - F(y'')\|^2 \\
&\geq (\tau/(L\|B\|)^2)\|B^T F(y') - B^T F(y'')\|^2 \\
&= \mu\|G(x') - G(x'')\|^2,
\end{aligned}$$

as desired. In this case the convergence of the projection methods follows from Theorem 13.2.

Let us now consider the Walrasian equilibrium model (12.8) with the cost mapping $G = -E$, so as E represents the excess demand, i.e.

$$E(p) = D(p) - S(p),$$

and $D(p)$ (respectively, $S(p)$) is the value of demand (respectively, supply) at the current price vector p. Suppose that each producer determines its supply as a solution of problem (12.5), then, by Proposition 12.2, its supply mapping is monotone on P, and so is the total supply mapping S, but it need not be strictly (strongly) monotone. Next, suppose that the negative demand mapping $-D$ is co-coercive, then the cost mapping $G(p) = -E(p) = S(p) - D(p)$ is clearly monotone, but may not be co-coercive. Hence, the usual projection (tâtonnement) process, defined by

$$p^{k+1} = \pi_P[p^k + \lambda_k E(p^k)], \lambda_k > 0,$$

does not ensure convergence in general. Nevertheless, Theorem 13.8 says that the splitting method, where the next iteration $p^{k+1} \in P$ satisfies the auxiliary VI:

$$\begin{aligned}(p - p^{k+1})^T[S(p^{k+1}) + \lambda^{-1}(p^{k+1} - p^k)\\ -D(p^k)] \geq 0 \quad \forall p \in P,\end{aligned} \tag{14.5}$$

generates the sequence $\{p^k\}$ converging to an equilibrium price vector. Observe that (14.5) may be interpreted as a modified version of the cobweb process and involves different velocities for supply and demand. It means that the price adjustment is slow for demand whereas it is immediate(rapid) for supply, but this dynamic process is stable. Conversely, in the same Walrasian model (12.8), we can apply the other splitting method, where $p^{k+1} \in P$ satisfies the auxiliary VI:

$$\begin{aligned}(p - p^{k+1})^T[S(p^k) + \lambda^{-1}(p^{k+1} - p^k) - D(p^{k+1})] \geq 0\\ \forall p \in P,\end{aligned}$$

which corresponds to the slow price reaction for supply and the rapid price reaction for demand and may be useful if the total supply of producers is at least co-coercive. Therefore, the slow price reaction requires strengthened monotonicity for ensuring the stability of the dynamic process.

14.2 Applications to systems of variational inequalities

Many equilibrium problems considered in Part I are formulated as extensions of the primal-dual system of VI's given in (4.11). In fact, this is the case for the extended Scarf model (5.5), (5.6), for the spatial price equilibrium models from Section 5.3, and for the multicommodity formulation of the network equilibrium problem from Section 6.1. Moreover, if we intend to remove complicated constraints determining the feasible set of VI, we can replace this VI by the primal -dual formulation, which leads to the system of VIs with simple constraints on variables. This approach can be applied to the linear programming problems (see (4.9)), to the Cassel-Wald model (see (5.1)), and to the migration equilibrium model from Section 6.2. The corresponding systems are also particular cases of (4.11). We now consider a somewhat more general primal-dual system of VIs presented in (11.26). The problem is to find a pair of points $(u^*, v^*) \in U \times V$ such that

$$\begin{array}{l} (u - u^*)^T[F(u^*) + \sum\limits_{i=1}^{m} v_i^* \nabla f_i(u^*)] \geq 0 \quad \forall u \in U, \\ (v - v^*)^T[b(v^*) - f(u^*)] \geq 0 \quad \forall v \in V, \end{array} \tag{14.6}$$

where $F : U \to \mathbb{R}^l$ and $b : V \to \mathbb{R}^m$ are continuous mappings, $f_i : \mathbb{R}^l \to \mathbb{R}, i = 1, \ldots, m$ are convex differentiable functions,

$$f(u) = (f_1(u), \ldots, f_m(u))^T,$$

$$U = \left\{u \in \mathbb{R}^l \mid u_j \geq 0 \quad \forall j \in J\right\}, \ J \subseteq \{1, \ldots, l\},$$

V is a nonempty, convex and closed subset of $\mathbb{R}^m$, for instance, we can set $V = \mathbb{R}^m_+$. Problem (14.6) covers the optimality conditions for convex programming problems, which correspond to the case when $b(v) \equiv 0$, $F(u) = \nabla f_0(u)$ for a convex function $f_0 : \mathbb{R}^l \to \mathbb{R}$ (see Proposition 11.5); the optimality conditions for constrained VIs, which correspond to the case when $b(v) \equiv 0$ (see Proposition 11.6); and the primal-dual system (4.11) in the case when $f_i, i = 1, \ldots, m$ are affine functions. At the same time, system (14.6) is equivalent to VI (14.1) where $x = (u, v), X = U \times V$,

$$G(x) = \begin{pmatrix} F(u) + \sum_{i=1}^m v_i \nabla f_i(u) \\ b(v) - f(u) \end{pmatrix}, \tag{14.7}$$

and we can apply the projection method (13.1) to find its solution. It can be specialized as follows. Given the current iterate $x^k = (u^k, v^k) \in U \times V$, the next iterate $x^{k+1} \in U \times V$ solves the auxiliary problem

$$\begin{array}{l} (u - u^{k+1})^T \left[F(u^k) + \sum\limits_{i=1}^{m} v_i^k \nabla f_i(u^k) \right. \\ \qquad \left. + \lambda_k^{-1}(u^{k+1} - u^k) \right] \geq 0 \quad \forall u \in U, \\ (v - v^{k+1})^T \left[b(v^k) - f(u^k) + \lambda_k^{-1}(v^{k+1} - v^k) \right] \geq 0 \quad \forall v \in V. \end{array} \tag{14.8}$$

In case $F(u) = \nabla f_0(u)$, $b(v) \equiv 0$, i.e. when (14.6) is the optimality conditions for the optimization problem (11.9)–(11.11), this method was proposed by *K.J. Arrow and L. Hurwicz* for finding saddle points of the Lagrange functions; see *Arrow, Hurwicz, and Uzawa* (1958), Chapter 6.

Exercise 14.5. Write down method (14.8) for system (4.11), for the spatial price equilibrium models from Section 5.3, and for the multicommodity formulation of the network equilibrium problem form Section 6.1.

If both the mappings F and b are strongly (strictly) monotone, then so is G in (14.7) due to Proposition 11.9 and, on account of Theorems 13.1, 13.3, 13.5, and 13.6, there exist rules for choice of λ_k ensuring convergence of the corresponding method to a solution of (14.6). Similarly, if F and b are co-coercive, then so is G and we can then apply Theorem 13.2 for providing convergence.

Exercise 14.6. Give an interpretation of (strong, strict) monotonicity and co-coercivity properties of F and b for the models indicated in Exercise 14.5.

Observe that the mapping G in (14.7) can not be integrable, hence Theorem 13.4 is not applicable. At the same time, we can replace the projection method with the splitting one (see (13.29)), which is explicit with respect to F and b and ensures convergence under the same condition on these mappings.

Exercise 14.7. Write down the indicated splitting method for system (14.6).

If even at least one of the mappings F or b does not possess strengthened monotonicity properties, we can not guarantee these properties for G and it becomes very difficult to obtain convergence of the projection method. But this is case for system (12.15) and for primal-dual systems which are optimality conditions for constrained problems, where we have $b(v) \equiv 0$; see *Gol'shtein and Tret'yakov* (1989), Chapter 6 for more details. However, using the method with different iterations in the variables u and v, we can obtain a suitable approach to finding solutions of the initial system (14.6).

Let us consider the case when $F : U \to \mathbb{R}^l$ is a strongly monotone mapping with constant $\tau > 0$, and $f : \mathbb{R}^l \to \mathbb{R}^m$ is a Lipschitz continuous mapping with constant L_f. Then, for each $v \in \mathbb{R}^m_+$, there exists the unique solution $u(v) \in U$ of the VI in the primal variables

$$(u - u(v))^T \left[F(u(v)) + \sum_{i=1}^{m} v_i \nabla f_i(u(v)) \right] \geq 0 \quad \forall u \in U. \tag{14.9}$$

Set

$$Q(v) = -f(u(v)),$$

thus defining the mapping $Q : \mathbb{R}^m_+ \to \mathbb{R}^m$.

Proposition 14.1. *Under the above assumptions the mapping Q is co-coercive with constant $\mu = \tau / L_f^2$.*

Proof. Fix $v', v'' \in \mathbb{R}^m_+$ and set $u' = u(v'), v'' = u(v'')$. Using (14.9), we have

$$(u'' - u')^T[F(u') + \sum_{i=1}^{m} v'_i \nabla f_i(u')] \geq 0$$

and

$$(u' - u'')^T[F(u'') + \sum_{i=1}^{m} v''_i \nabla f_i(u'')] \geq 0.$$

Adding these inequalities gives

$$\begin{aligned}&\sum_{i=1}^{m} v'_i (u'' - u')^T \nabla f_i(u') + \sum_{i=1}^{m} v''_i (u' - u'')^T \nabla f_i(u'')\\ &\geq (u'' - u')^T[F(u'') - F(u')]\\ &\geq \tau \|u'' - u'\|^2.\end{aligned}$$

Since the functions $f_i, i = 1, \ldots, m$, are convex, using Proposition 11.4 now yields

$$\begin{aligned}&(v'' - v')^T[Q(v'') - Q(v')]\\ &= \sum_{i=1}^{m} v'_i[f_i(u'') - f_i(u')] + \sum_{i=1}^{m} v''_i[f_i(u') - f_i(u'')]\\ &\geq \tau \|u'' - u'\|^2 \geq (\tau / L_f^2)\|f(u'') - f(u')\|^2\\ &= \mu \|Q(v'') - Q(v')\|^2,\end{aligned}$$

i.e. Q is co-coercive, as desired. □

Observe that system (14.6) is now equivalent to the following VI in the dual variables: Fix $v^* \in V$ such that

$$(v - v^*)^T[b(v^*) + Q(v^*)] \geq 0 \quad \forall v \in V, \tag{14.10}$$

as the following proposition states.

Proposition 14.2. *Let the above assumptions hold. If (u^*, v^*) solves (14.6), then v^* solves (14.10) and $u^* = u(v^*)$. Conversely, if v^* solves (14.10), then the pair (u^*, v^*), where $u^* = u(v^*)$ solves (14.6).*

Exercise 14.8. Prove the assertion of Proposition 14.2.

The projection method (13.1) applied to (14.10) consists in finding the next iterate $v^{k+1} \in V$ from the VI

$$(v - v^{k+1})^T[b(v^k) + Q(v^k) + \lambda_k^{-1}(v^{k+1} - v^k)] \geq 0 \quad \forall v \in V, \tag{14.11}$$

In case $F(u) = \nabla f_0(u)$, $b(v) \equiv 0$, which corresponds to the optimality conditions for optimization problems, this dual method was proposed by *H. Uzawa*; see *Arrow, Hurwicz, and Uzawa* (1958), Chapter 7. We now give a convergence result under the co-coercivity of b, note that this is the case if b is a constant mapping.

Theorem 14.1. *Suppose that all the above assumptions are fulfilled, system (14.6) is solvable, and $b : \mathbb{R}^m_+ \to \mathbb{R}^m$ is co-coercive with constant $\tilde{\mu} > 0$. Then any sequence $\{v^k\}$, generated by method (14.11) with $\lambda_k = \lambda \in (0, 2\mu')$, $\mu' = \min\{\mu, \tilde{\mu}\}$, converges to a point $v^* \in V$, and the sequence $\{u^k\}$, with $u^k = u(v^k)$, converges to a point $u^* \in U$ such that u^* solves (14.10) and (u^*, v^*) solves (14.6).*

Proof. For brevity, set $H(v) = b(v) + Q(v)$. Then, for an arbitrary pair of points $v', v'' \in V$, we have

$$\begin{aligned}
&(v' - v'')^T[H(v') - H(v'')] \\
&\geq \tilde{\mu}\|b(v') - b(v'')\|^2 + \mu\|Q(v') - Q(v'')\|^2 \\
&\geq \mu'\|H(v') - H(v'')\|^2,
\end{aligned}$$

i.e. H is co-coercive with constant μ'. On account of Theorem 13.2, $\{v^k\}$ converges to a point v^* which is a solution of VI (14.10). Since Q is co-coercive, it is continuous, hence so is the mapping $v \mapsto u(v)$. Therefore, the sequence $\{u^k\}$ with $u^k = u(v^k)$ has limit points, but all these points coincide with $u^* = u(v^*)$, hence $\lim_{k\to\infty} u^k = u^*$. By Proposition 14.2, (u^*, v^*) is a solution of (14.6). □

It has been noticed that the equilibrium model (4.11), the extended Scarf model (5.5)–(5.6), and system (12.15) are particular cases of system (14.6). If we choose y as the dual variable in (4.11), then method (14.11) will correspond to the tâtonnement process due to (5.7) (cf. (10.4)). This is also the case for systems (5.5)–(5.6), and (12.15) if we choose the prices p as dual variables.

Exercise 14.9. Write down method (14.11) for systems (4.11), (5.5)–(5.6), and (12.15) and establish its convergence properties.

By using the other theorems from Chapter 13, we can obtain several convergence results for method (14.11) under the other assumptions on b.

Exercise 14.10. Prove similar convergence results for method (14.11) for the cases when b is strongly monotone and strictly monotone.

We can somewhat weaken the assumptions for convergence by using the splitting method (13.29). This method applied to (14.10) consists in finding the next iterate $v^{k+1} \in V$ from the VI

$$\begin{array}{l}(v - v^{k+1})^T[b(v^{k+1}) \\ +\lambda_k^{-1}(v^{k+1} - v^k) + Q(v^k)] \geq 0 \quad \forall v \in V,\end{array} \tag{14.12}$$

i.e. it is implicit with respect to b and explicit with respect to Q. If $b(v) \equiv b$, (14.12) clearly coincides with (14.11). Combining Theorem 13.8 and Propositions 14.1 and 14.2, we can obtain the convergence theorem for this method.

Exercise 14.11. Write down method (14.12) for systems (4.11), (5.5)–(5.6), and (12.15).

Theorem 14.2. *Suppose that all the assumptions above are fulfilled, system (14.6) is solvable, and $b : \mathbb{R}^m_+ \to \mathbb{R}^m$ is monotone. Then any sequence $\{v^k\}$ generated by the method (14.12) with $\lambda_k = \lambda \in (0, 2\mu)$, converges to a point $v^* \in V$, and the sequence $\{u^k\}$, with $u^k = u(v^k)$, converges to a point $u^* \in U$ such that v^* solves (14.10), and (u^*, v^*) solves (14.6).*

Exercise 14.12. Prove the assertion of Theorem 14.2.

Chapter 15

Regularization Methods

In was noticed in Chapter 13 that the projection method (13.1) does not ensure convergence to a solution of VI in the case when its cost mapping is only monotone, but does not possess strengthened monotonicity properties; see Example 13.1 and Exercise 13.3. At the same time, these strengthened properties do not hold in many equilibrium models; see Chapter 12. Moreover, the corresponding cost mappings are not integrable, thus preventing the application of the descent approach described in Theorem 13.4. Therefore, we should present iterative methods for the general monotone case, which can be also regarded as more stable dynamic processes in comparison with those in Chapters 13–14.

Throughout this and the next chapter, we shall consider the following VI: Find a point $x^* \in X$ such that

$$(x - x^*)^T G(x^*) \geq 0 \quad \forall x \in X, \tag{15.1}$$

where X is a nonempty, convex, and closed subset in $\mathbb{R}^n$, $G : X \to \mathbb{R}^n$ is a continuous monotone mapping. As before, X^* denotes the solution set of this problem. First we consider regularization type methods which replace the initial monotone VI (15.1) with a sequence of auxiliary VI's with strengthened monotonicity properties.

15.1 The classical regularization method and its modifications

The most popular and simple *regularization method* was proposed by *A.N. Tikhonov*; see *Tikhonov and Arsenin* (1977). This method consists in replacing VI (15.1) by a sequence of auxiliary VI's of the form: Find $x^\varepsilon \in X$ such that

$$(x - x^\varepsilon)^T [G(x^\varepsilon) + \varepsilon x^\varepsilon] \geq 0 \quad \forall x \in X, \tag{15.2}$$

where $\varepsilon > 0$ is a parameter. Since G is monotone, $G + \varepsilon I$ is strongly monotone and, by Theorem 11.4, (15.2) has a unique solution, which can be found by one of the versions of the projection method from Chapter 13 within a given accuracy. The basic approximation property of the sequence $\{x^\varepsilon\}$ is formulated as follows.

Theorem 15.1. *If VI (15.1) is solvable, the sequence $\{x^{\varepsilon_k}\}$ obtained from (15.2) with $\{\varepsilon_k\} \to 0$ converges to the solution x_n^* of (15.1) nearest to origin.*

Proof. Take an arbitrary solution x^* of VI (15.1). Then, by definition, we have

$$(x^\varepsilon - x^*)^T G(x^\varepsilon) \geq (x^\varepsilon - x^*)^T G(x^*) \geq 0$$

and

$$(x^* - x^\varepsilon)^T [G(x^\varepsilon) + \varepsilon x^\varepsilon] \geq 0.$$

Adding these inequalities gives $(x^*)^T x^\varepsilon \geq \|x^\varepsilon\|^2$, i.e.

$$\|x^*\| \geq \|x^\varepsilon\| \quad \forall x^* \in X^*. \tag{15.3}$$

It follows that the sequence $\{x^{\varepsilon_k}\}$ is bounded, hence it has limit points. Taking the limit $k \to \infty$ in (15.2), we obtain that each limit point of $\{x^{\varepsilon_k}\}$ belongs to X^*. At the same time, the monotonicity of G implies that, for all $x \in X$, we have

$$(x - x^*)^T G(x) \geq 0 \quad \forall x^* \in X^*.$$

Hence X^* is convex and closed, and, by Proposition 11.12, there exists the unique projection x_n^* of origin onto X^*. Applying (15.3) now gives

$$\|x_n^*\| \geq \|x'\|, \quad x' \in X^*$$

for an arbitrary limit point x' of $\{x^{\varepsilon_k}\}$, i.e. all these limit points coincide with x_n^*. □

Thus, combining the approximate solution of perturbed problems (15.2), which possess the strong monotonicity, with driving the regularization parameter ε to zero, we can obtain a solution of the initial monotone VI. Therefore, the dynamic processes with reactions on the perturbed mapping appears more stable in comparison with the processes without such perturbations.

The regularization approach admits various modifications. One of them was proposed by *A.B. Bakushinskii and B.T. Polyak* and is called the *iterative regularization.* The idea of this approach consists in simultaneous changes of the regularization parameters and the stepsizes of an approximation method. If we take the projection method (13.1) as a basis, the corresponding iterative procedure can be described as follows:

$$x^{k+1} = \pi_X(x^k - \lambda_k[G(x^k) + \varepsilon_k x^k]), \quad \varepsilon_k > 0, \ \lambda_k > 0; \tag{15.4}$$

where

$$\begin{aligned} &\lim_{k\to\infty} \varepsilon_k = 0, \ \lim_{k\to\infty} (\lambda_k/\varepsilon_k) = 0, \\ &\lim_{k\to\infty} \frac{\varepsilon_k - \varepsilon_{k+1}}{\lambda_k \varepsilon_k^2} = 0, \ \sum_{k=0}^{\infty} (\varepsilon_k \lambda_k) = \infty. \end{aligned} \tag{15.5}$$

Proposition 15.1. *(see Bakushinskii and Goncharskii (1989), Theorem 3.1) If there exists a constant L such that*

$$\|G(x)\| \leq L(1 + \|x\|) \quad \forall x \in X,$$

then any sequence $\{x^k\}$ generated in conformity with rules (15.4) - (15.5) converges to the point x_n^*.

Observe that the conditions in (15.5) are fulfilled if we set

$$\lambda_k = (k+1)^{-1/2}, \varepsilon_k = (k+1)^{-\tau}, \tau \in (0, 1/2).$$

15.2 The proximal point method

The *proximal point method* represents some other way of constructing a sequence of perturbed problems, where the regularization parameter may be fixed. This method was first proposed by *B. Martinet.* Observe that the projection method does not ensure convergence for the initial monotone VI (15.1), hence its convergence may be very slow if it is applied to the perturbed VI (15.2) where ε is small enough, but tending ε to zero is necessary for convergence of the regularization method. Therefore, the fixed regularization parameter yields certain advantages.

The proximal mapping has been considered in Section 13.3. We now recall its properties with respect to G, i.e. we consider the case when $F \equiv 0$ in (13.27). Given a point $x \in \mathbb{R}^n$, there exists a unique solution of the following VI: Find $P(x) \in X$ such

$$(y - P(x))^T \left(G[P(x)] + \lambda^{-1}[P(x) - x]\right) \geq 0 \quad \forall y \in X. \tag{15.6}$$

where $\lambda > 0$ is a fixed number, since G is monotone and continuous. The proximal method consists in generating a sequence $\{x^k\}$ as follows:

$$x^{k+1} = P(x^k). \tag{15.7}$$

Its convergence properties can be deduced from Theorem 13.8, but we give some other proof for the clarity of exposition. First we recall the properties of the proximal mapping obtained in Propositions 13.3 and 13.4.

Lemma 15.1. *(i) The mapping P is non-expansive and continuous and*

$$(x'' - x')^T [P(x') - P(x'')] \geq \|P(x') - P(x'')\|^2 \quad \forall x', x'' \in \mathbb{R}^n.$$

(ii) A point $x^* \in X$ is a solution of VI (15.1) if and only if $x^* = P(x^*)$.

For each point $x \in \mathbb{R}^n$, set

$$H(x) = \lambda^{-1}[x - P(x)],$$

thus defining the mapping $H : \mathbb{R}^n \to \mathbb{R}^n$. We intend to show that H is co-coercive with constant λ.

Lemma 15.2. *For each pair of points $x', x'' \in \mathbb{R}^n$, it holds that*

$$(x'' - x')^T [H(x'') - H(x')] \geq \lambda \|H(x'') - H(x')\|^2.$$

Proof. Fix x', x''. Then, using Lemma 15.1 (i), we have

$$\begin{aligned}
&(x'' - x')^T [H(x'') - H(x')] \\
&= \lambda [H(x'') - H(x')]^T [H(x'') - H(x')] \\
&+ [P(x'') - P(x')]^T [H(x'') - H(x')] \\
&= \lambda \|H(x'') - H(x')\|^2 \\
&+ \lambda^{-1} \left[(x'' - x')^T [P(x'') - P(x')] - \|P(x'') - P(x')\|^2 \right] \\
&\geq \lambda \|H(x'') - H(x')\|^2,
\end{aligned}$$

and the result follows. □

We now observe that (15.7) coincides with the simple iteration

$$x^{k+1} = x^k - \lambda H(x^k) \tag{15.8}$$

applied to finding a solution of the equation

$$H(x^*) = 0. \tag{15.9}$$

Theorem 15.2. *Any sequence $\{x^k\}$ satisfying (15.7) converges to a solution of VI (15.1) for an arbitrary $\lambda > 0$.*

Proof. If $\{x^k\}$ is generated in conformity with (15.7), we obtain (15.8) by definition. Taking into account Lemma 15.2, we conclude that $\{x^k\}$ converges to a solution x^* of equation (15.9) due to Theorem 13.2. It follows now from Lemma 15.1 (ii) that x^* is a solution of VI (15.1). □

Thus, the proximal point method involves a solution of the auxiliary strongly monotone VI (15.6) for $x = x^k$ at each iteration. The exact solution of this problem may be too hard in implementation but it then can be replaced by an inexact solution within certain prescribed accuracy. Such inexact versions maintain the basic convergence properties; see *Gol'shtein and Tret'yakov* (1989) and *Facchinei and Pang* (2003) for more details.

Chapter 16

Direct Iterative Methods for Monotone Variational Inequalities

In this chapter, we also consider iterative solution methods for VI (15.1) with the monotone and continuous cost mapping G. These methods are based on modifying direction finding procedures, i.e. unlike the regularization methods, they do not utilize perturbations of the initial problem.

16.1 Extrapolation methods

It was noticed in Chapter 13 that the projection method, which represents a dynamic process with reactions on the current state of the system, does not ensure convergence to a solution of VI (15.1) under the above general assumptions. In particular, this is the case for its specialization (14.8), which is known as the Arrow-Hurwicz method, when F and b in (14.6) are only monotone, see Example 13.1 and Exercise 13.3. Clearly, this is also the case for the primal-dual system (4.9) related to linear programming problems. However, if we replace method (14.8) with the reaction on the current state by a method with the reaction on the extrapolated state, then it can ensure convergence to a solution. This property was noticed by *T. Kose* and modified and extended by *K.J. Arrow and R. Solow*; see *Arrow, Hurwicz, and Uzawa* (1958), Chapter 11, for more details. For the general VI of form (15.1) such a method was proposed by *G.M. Korpelevich* and it is called the *extragradient method*. Its iteration is as follows:

$$\begin{aligned} x^{k+1} &= \pi_X[x^k - \lambda_k G(y^k)], \\ y^k &= \pi_X[x^k - \lambda_k G(x^k)], \lambda_k > 0. \end{aligned} \tag{16.1}$$

The convergence property of method (16.1) for the monotone VI (15.1) is based on the fact that the angle between $-G(y^k)$ and $x^* - x^k$ can be made acute for any $x^* \in X^*$, so that the iteration process (16.1), under a suitable choice of the stepsize λ_k, becomes convergent to a solution of VI (15.1). Observe that the extragradient method involves the same stepsize for both steps in (16.1). For instance, we can set $\lambda_k = \lambda \in (0, 1/L)$, where L is the Lipschitz constant for G or utilize a linesearch procedure: see *Gol'shtein and Tret'yakov* (1989), Chapter 5 and *Facchinei and Pang* (2003), Chapter 12 for more details.

We now consider another way of providing convergence via extrapolation steps, which was proposed by the author of this book and called the *combined relaxation method.* In this method, the first part of each iteration serves for computing parameters of a hyperplane separating strictly the current iterate and the solution set X^*, whereas the second part consists in making the projections on this hyperplane and on the feasible set, if necessary. As a result, we obtain the monotone decrease of the distance to all solutions. The first part of the iteration may be in principle based on an iteration on most relaxation methods, but now we present the simplest projection-based iteration which is close to the extragradient method; see *Konnov* (2001), *Facchinei and Pang* (2003), and references therein.

Algorithm (feasible combined relaxation). Choose a point $x^0 \in X$, numbers $\alpha \in (0,1)$, $\beta \in (0,1)$, $\gamma \in (0,2)$, $\lambda > 0$, and set $k = 0$.

Step 1 (Auxiliary procedure): Compute $z^k = \pi_X[x^k - G(x^k)]$ and set $p^k = z^k - x^k$. If $p^k = 0$, stop. Otherwise determine m as the smallest non-negative integer such that

$$(p^k)^T G(x^k + \beta^m p^k) \leq \alpha (p^k)^T G(x^k), \tag{16.2}$$

set $\theta_k = \beta^m$, $y^k = x^k + \theta_k p^k$. If $G(y^k) = 0$, stop.

Step 2 (Main iteration): Set

$$\begin{aligned} g^k &= G(y^k), \omega_k = (g^k)^T(x^k - y^k), \\ x^{k+1} &= \pi_X[x^k - \gamma(\omega_k/\|g^k\|^2)g^k], \end{aligned} \tag{16.3}$$

$k = k + 1$ and go to Step 1.

According to the description, the algorithm finds a solution to VI in the case of its finite termination. Therefore, in what follows we shall consider only the case of the infinite sequence $\{x^k\}$.

The basic properties of the algorithm are collected in the next lemma.

Lemma 16.1. *(i) The linesearch procedure in Step 1 is always finite.*

(ii) It holds that

$$(g^k)^T(x^k - x^*) \geq \omega_k \geq \alpha\theta_k\lambda^{-1}\|p^k\|^2 > 0 \tag{16.4}$$

and

$$\|x^{k+1} - x^*\|^2 \leq \|x^k - x^*\|^2 - \gamma(2-\gamma)(\omega_k/\|g^k\|)^2 \tag{16.5}$$

for each $x^* \in X^*$.

Proof. If we suppose that the linesearch procedure is infinite, then (16.2) holds for $m \to \infty$, hence, by continuity of G,

$$(1-\alpha)(z^k - x^k)^T G(x^k) \leq 0.$$

On the other hand, by Proposition 11.12,

$$(y - z^k)^T \left[G(x^k) + \lambda^{-1}(z^k - x^k)\right] \geq 0 \quad \forall y \in X, \tag{16.6}$$

hence

$$(x^k - z^k)^T \left[G(x^k) + \lambda^{-1}(z^k - x^k)\right] \geq 0,$$

and

$$(x^k - z^k)^T G(x^k) \geq \lambda^{-1}\|x^k - z^k\|^2. \tag{16.7}$$

It now follows that $x^k = z^k$, which contradicts the construction of the algorithm. Hence, (i) is true.

Next, by using (16.2), (16.3), and (16.7) and the monotonicity of G, we have

$$\begin{array}{l} (g^k)^T(x^k - x^*) \\ = (x^k - y^k)^T G(y^k) + (y^k - x^*)^T G(y^k) \\ = \omega_k + (y^k - x^*)^T G(y^k) \geq \omega_k \\ = \theta_k(x^k - z^k)^T G(y^k) \\ \geq \alpha\theta_k(x^k - z^k)^T G(x^k) \geq \alpha\theta_k\lambda^{-1}\|x^k - z^k\|^2, \end{array} \tag{16.8}$$

i.e. (16.4) is also true. By (16.4) and the projection properties, we have

$$\begin{array}{l} \|x^{k+1} - x^*\|^2 \leq \|x^k - \gamma(\omega_k/\|g^k\|^2)g^k - x^*\|^2 \\ = \|x^k - x^*\|^2 - 2\gamma(\omega_k/\|g^k\|^2)(g^k)^T(x^k - x^*) \\ +(\gamma_k\omega_k/\|g^k\|)^2 \\ \leq \|x^k - x^*\|^2 - 2\gamma(2-\gamma)(\omega_k/\|g^k\|)^2, \end{array}$$

i.e. (16.5) is fulfilled, as desired. □

Figure 16.1:

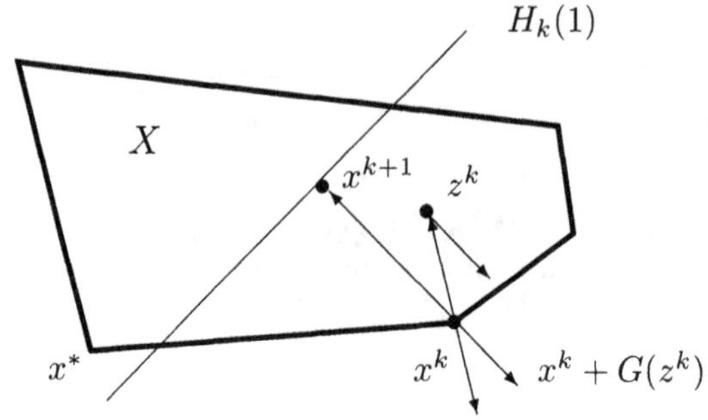

Thus, the above algorithm is well-defined and relation (16.4) says that the vector g^k and the number $\omega_k > 0$ determines the family of hyperplanes

$$H_k(\gamma) = \{y \in \mathbb{R}^n \mid (g^k)^T(x^k - y) = \gamma\omega_k\}$$

such that $H_k(1)$ separates x^k and X^*. Moreover, the point $\tilde{x}^{k+1} = x^k - \gamma(\omega_k/\|g^k\|^2)g^k$ is nothing but the projection of x^k onto $H_k(\gamma)$, hence the distance from $\tilde{x}^{k+1}$ to each point X^* decreases and the same assertion is true for x^{k+1} due to (16.5); see Figure 16.1. This result yields immediately very useful properties of $\{x^k\}$.

Lemma 16.2. *Suppose that a sequence $\{x^k\}$ is generated by the above algorithm. Then:*

(i) $\{x^k\}$ is bounded.

(ii) $\sum\limits_{k=0}^{\infty}(\omega_k/\|g^k\|)^2 < \infty$.

(iii) For each limit point x^* of $\{x^k\}$ such that $x^* \in X^*$ we have

$$\lim_{k\to\infty} x^k = x^*.$$

We are now ready to obtain convergence of the algorithm.

Theorem 16.1. *If a sequence $\{x^k\}$ is generated by the above algorithm, then*

$$\lim_{k\to\infty} x^k = x^* \in X^*.$$

Proof. By Lemma 16.2 (i), $\{x^k\}$ is bounded, hence so are $\{z^k\}$ and $\{y^k\}$ because of (16.7). Let us consider two possible cases.
Case 1: $\lim\limits_{k\to\infty} \theta_k = 0$.
Set $\tilde{y}^k = x^k + (\theta_k/\beta)p^k$, then $(p^k)^T G(\tilde{y}^k) > \alpha(p^k)^T G(x^k)$. Select convergent subsequences $\{x^{k_q}\} \to x'$ and $\{z^{k_q}\} \to z'$, then $\{\tilde{y}^{k_q}\} \to x'$ since $\{x^k\}$ and $\{z^k\}$ are bounded. By continuity, we have

$$(1-\alpha)(z'-x')^T G(x') \geq 0,$$

but taking the same limit in (16.7) gives

$$(x'-z')^T G(x') \geq \lambda^{-1}\|x'-z'\|^2,$$

i.e., $x' = z'$ and (16.6) now yields

$$(y-x')^T G(x') \geq 0 \quad \forall y \in X, \tag{16.9}$$

i.e., $x' \in X^*$.
Case 2: $\limsup\limits_{k\to\infty} \theta_k \geq \tilde{\theta} > 0$.
It means that there exists a subsequence $\{\theta_{k_q}\}$ such that $\theta_{k_q} \geq \tilde{\theta} > 0$. Combining this property with Lemma 16.2 (ii) and (16.8) gives

$$\lim_{q\to\infty} \|x^{k_q} - z^{k_q}\| = 0.$$

Without loss of generality we can suppose that $\{x^{k_q}\} \to x'$ and $\{z^{k_q}\} \to z'$, then $x' = z'$. Again, taking the corresponding limit in (16.6) yields (16.9), i.e. $x' \in X^*$. The assertion of the theorem follows from Lemma 16.2 (iii). □

Combined relaxation methods may utilize iterations of various algorithms as bases of the auxiliary procedure for computing the parameters of the separating hyperplane and different rules for determining this hyperplane. To illustrate this assertion, we present now some other combined relaxation method.

Algorithm (infeasible combined relaxation). Choose a point $x^0 \in \mathbb{R}^n$, numbers $\alpha \in (0,1)$, $\beta \in (0,1)$, $\gamma \in (0,2)$, $\lambda > 0$, and a sequence of of mappings $\{F^{(k)} : \mathbb{R}^n \to \mathbb{R}^n\}$ such that each $F^{(k)}$ is strongly monotone with constant $\tau' > 0$ and Lipschitz continuous with constant $\tau'' > 0$ for $k = 0, 1, \ldots$ Set $k = 0$.

Step 1 (Auxiliary procedure): Find m as the smallest non-negative integer such that

$$\begin{aligned}&(x^k - z^{k,m})^T [G(x^k) - G(z^{k,m})] \\ &\leq (1-\alpha)(\lambda\beta^m)^{-1}(x^k - z^{k,m})^T \left[F^{(k)}(x^k) - F^{(k)}(z^{k,m})\right],\end{aligned}$$

where $z^{k,m}$ is a solution of the auxiliary problem

$$(y - z^{k,m})^T \left(G(x^k) + (\lambda\beta^m)^{-1}\left[F^{(k)}(z^{k,m}) - F^{(k)}(x^k)\right]\right) \geq 0 \quad \forall y \in X.$$

Set $\theta_k = \beta^m\lambda$, $y^k = z^{k,m}$. If $x^k = y^k$ or $G(y^k) = 0$, stop.

Step 2 (Main iteration): Set

$$\begin{aligned} g^k &= G(y^k) - G(x^k) - \theta_k^{-1}\left[F^{(k)}(y^k) - F^{(k)}(x^k)\right], \\ \omega_k &= (g^k)^T(x^k - y^k), \\ x^{k+1} &= x^k - \gamma(\omega_k/\|g^k\|^2)g^k, \end{aligned}$$

$k = k + 1$ and go to Step 1.

Observe that we can set for example $F^{(k)}(x) = A_k x$, where A_k is a positive definite matrix and the choice $A_k \equiv I$ in this algorithm leads to the usual auxiliary projection iteration, but the algorithm utilizes some other linesearch strategy and the rule for determining the parameters g^k and ω_k. Also, it involves the assumption that G is defined on the whole space $\mathbb{R}^n$, i.e. $G : \mathbb{R}^n \to \mathbb{R}^n$, then the iteration sequence $\{x^k\}$ is infeasible. Nevertheless, the parameters g^k and ω_k also ensure the crucial property that $H_k(1)$ separates x^k and X^*. As a result, the above algorithm has similar convergence properties. This approach can be extended for various classes of VIs, including multi-valued and generalized monotone ones; see *Konnov* (2001) for more details.

16.2 The ellipsoid method

We also consider VI (15.1) with the monotone and continuous cost mapping G. The idea of the *ellipsoid method* consists in construction of a sequence of ellipsoids $\{E_k\}$, such that each E_k contains a point of X^*, with $\text{vol}(E_k)$ tending to zero as $k \to \infty$. Here $\text{vol}(E_k)$ denotes the volume of E_k. The implementation of the method is based on the following observations.

Let z be a point in $\mathbb{R}^n$ and let A be an $n \times n$ positive definite matrix. Then one can define the ellipsoid $\text{ell}(A, z)$ as follows:

$$\text{ell}(A, z) = \{x \in \mathbb{R}^n \mid (x - z)^T A^{-1}(x - z) \leq 1\}.$$

It is clear that $\text{ell}(I, 0) = \{x \in \mathbb{R}^n \mid \|x\| \leq 1\}$.

Proposition 16.1. *(Schrijver (1986), Theorem 13.1) The ellipsoid $ell(A', z')$, where*

$$\begin{aligned} z' &= z - \frac{1}{n+1} \frac{Aq}{\sqrt{q^T Aq}}, \\ A' &= \frac{n^2}{n^2 - 1} \left(A - \frac{2}{n+1} \frac{Aqq^T A^T}{q^T Aq} \right), \end{aligned} \tag{16.10}$$

has the minimal volume among all those containing the half-ellipsoid

$$ell(A, z) \bigcap \{x \in \mathbb{R}^n \mid (x - z)^T q \leq 0\}.$$

Moreover,

$$\frac{vol[\ ell(A', z')]}{vol[\ ell(A, z)]} = \left(\frac{n^2}{n^2 - 1} \right)^{\frac{n-1}{2}} \times \left(\frac{n}{n+1} \right) < e^{-\frac{1}{2(n+1)}}. \tag{16.11}$$

Thus, the volume of the ellipsoid $\text{ell}(A', z')$ will reduce if we make use of (16.10) to find the next ellipsoid. Next, given a point $z \in \mathbb{R}^n$, we can choose the element q in (16.10) as follows

$$q \in Q(z) = \begin{cases} \{G(z)\} & \text{if } z \in X, \\ \{p \in \mathbb{R}^n \mid (y - z)^T p \leq 0 \quad \forall y \in X\} & \text{if } z \notin X. \end{cases}$$

If $\text{ell}(A, z)$ contains a point $x^* \in X$, then

$$x^* \in \{x \in \mathbb{R}^n \mid (x - z)^T q \leq 0\}$$

hence $x^* \in \text{ell}(A', z')$. Taking these properties as a basis, we can construct an iteration sequence.

Algorithm (ellipsoid). Choose a point $x^0 \in \mathbb{R}^n$, a number $\lambda > 0$ such that $\|x^0 - x^*\| \leq \lambda$ for some $x^* \in X^*$ and set $A_0 = \lambda^2 I$. At the k-th iteration, $k = 0, 1, \ldots$, choose $q^k \in Q(x^k)$, set

$$\begin{aligned} x^{k+1} &= x^k - \frac{1}{n+1} \frac{A_k q^k}{\sqrt{(q^k)^T A_k q^k}}, \\ A_{k+1} &= \frac{n^2}{n^2 - 1} \left(A_k - \frac{2}{n+1} \frac{A_k q^k (A_k q^k)^T}{(q^k)^T A_k q^k} \right), \end{aligned}$$

and $k = k + 1$.

To justify the above algorithm, it suffices to observe that the recurrence (16.10) preserves the positive definiteness of A_{k+1}. By description, $E_k =$ ell(A_k, x^k). Therefore, we must have $x^* \in$ ell(A_k, x^k), and, by (16.11), vol (ell(A_k, x^k)) tends to zero in a linear rate. These properties yield the convergence of the sequence $\{x^k\}$ to a solution.

Exercise 16.1. Find several first steps of the ellipsoid algorithm applied to VI (15.1) where $n = 2$, $G(x) = (x_2, -x_1)^T$,

$$X = \{x \in \mathbb{R}^2 \mid x_2 \geq 0, x_1 - x_2^2 \geq 0\},$$

utilizing the starting values $x^0 = (1, 0)^T$, $\lambda = 2$ and give their graphical illustration.

Also, the ellipsoid algorithm admits extensions for multi-valued and monotone VIs. It was first proposed by *D.B. Yudin and A.S. Nemirovskii* and by *N.Z. Shor* for convex optimization problems. We refer to the books by *Polyak* (1983), *Schrijver* (1986), and *Konnov* (2001) and to the bibliography therein for a more detailed description of the properties of this method.

Chapter 17

Solutions to Exercises

We consider here mostly exercises whose solution may cause certain theoretical difficulties.

Chapter 1

Exercise 1.1 *Prove the equivalence of (1.7) and (1.8).*

Let p^* solve (1.7). Then, for each $p \in [p', p'']$, we have

$$(p^* - p)\begin{cases} \leq 0 & \text{if } p^* = p', \\ \geq 0 & \text{if } p^* = p''. \end{cases}$$

Using (1.7), we see that

$$E(p^*)(p^* - p) \geq 0,$$

i.e. (1.8) holds. Conversely, let p^* solve (1.8). Then $p^* \in (p', p'')$ implies $E(p^*) = 0$, similarly, $p^* = p'$ ($p^* = p''$) implies $E(p^*) \leq 0$ ($E(p^*) \geq 0$) and (1.7) holds. □

Exercise 1.2 *Suppose that E is continuous and prove that (1.7) is solvable if $-\infty < p' \leq p'' < +\infty$. Find sufficient solvability conditions for (1.7) if either $p' = -\infty$ or $p'' = +\infty$.*

If either $E(p') \leq 0$ or $E(p'') \geq 0$, then $p^* = p'$ (respectively, $p^* = p''$). Otherwise we have $E(p') > 0$ and $E(p'') < 0$. Since E is continuous, applying the mean value theorem we see that there exists $p^* \in (p', p'')$ such that $E(p^*) = 0$.

Next, if $p' = -\infty$ and there exists $\tilde{p} \leq p''$ such that $E(\tilde{p}) \geq 0$, then (1.7) is solvable. In fact, $E(p'') \geq 0$ implies that p'' solves (1.7). Otherwise we have $E(p'') < 0$ and by the mean value theorem, there exists $p^* \in [\tilde{p}, p'']$ such that $E(p^*) = 0$. Similarly, if $p'' = +\infty$ and there exists $\tilde{p} \geq p'$ such that $E(\tilde{p}) \leq 0$, then (1.7) is solvable. □

Chapter 2

Exercise 2.2 *Prove the assertion of Theorem 2.3 under condition (2.6).*

Replacing A by its transpose A^T in the proof of Theorem 2.3 and utilizing (2.6) instead of (2.7), we obtain $\lambda_{A^T} < 1$, but the matrices A and A^T have the same set of eigenvalues, hence $\lambda_A < 1$ and the result follows. □

Exercise 2.3 *Prove the inequality $\lambda_A < 1$ in Theorem 2.5.*

By Theorem 2.4, x_A is now positive. Hence, following the proof of Theorem 2.3, we see that the crucial inequality (2.8) holds in case (2.7′). Similarly, we should replace A by its transpose A^T in case (2.6′). □

Exercise 2.4 *Prove that an $n \times n$ matrix A is indecomposable if, for each pair (i, j), $i \neq j$, there exist indices $i_1, i_2, \ldots, i_{s-1}$ such that $a_{i_{l-1}i_l} > 0$ for $l = 1, 2, \ldots, s$, where $i_0 = i$, $i_s = j$.*

We prove this assertion by contradiction. If A is decomposable, then there exists an index set $J \subset \{1, \ldots, n\}$ such that $a_{kl} = 0$ as $k \notin J$, $l \in J$. If we take $i_0 \notin J, i_s \in J$, then there exists a pair (i_{l-1}, i_l) such that $a_{i_{l-1}i_l} > 0$, but $i_{l-1} \notin J$, $i_l \in J$, and the result follows. □

Exercise 2.5 *Write down the extension of the input-output model for the case (2.2′).*

The problem is to find $x^* \in [x', x'']$ such that

$$(x - x^*)^T \left[(I - A)x^* - y\right] \geq 0 \quad \forall x \in [x', x''].$$

In fact, if x^* solves this problem and $x' < x^* < x''$, then $(I - A)x^* - y = 0$ (cf. (2.2)). The more detailed consideration is given in Sections 7.1 and 8.1. □

Exercise 2.6 *Prove that the solution set of problem (2.12) is a convex cone.*

If π solves (2.12) (or equivalently, (2.12′)), then so is $\lambda\pi$ for any $\lambda \geq 0$, hence the solution set is a cone. Next, if π' and π'' are solutions of (2.12) and $\alpha \in (0,1)$, then

$$(I-A)\left(\alpha\pi' + (1-\alpha)\pi''\right)$$
$$= \alpha(I-A)\pi' + (1-\alpha)(I-A)\pi'' = 0$$

and

$$\alpha\pi' + (1-\alpha)\pi'' \geq 0,$$

i.e. $\alpha\pi' + (1-\alpha)\pi''$ also solves (2.12), and the solution set is a convex cone. □

Chapter 3

Exercise 3.1 *Prove that the indecomposability of the set (A, I) is equivalent to the indecomposability of the non-negative matrix A.*

By definition, the set (A, I) is indecomposable, if there are no nonempty index subsets $K \subsetneq \{1, \ldots, n\}$ and $J \subsetneq \{1, \ldots, n\}$ such that

$$\sum_{j \in J} a_{ij} = 0 \quad \text{for each } i \notin K,$$

where $K \bigcap J = \emptyset$ since $B = I$. But this is equivalent to the indecomposability of the matrix A. □

Chapter 4

Exercise 4.1 *Prove the assertions of Theorems 4.1 and 4.2.*

Combining (4.7) and (4.5) gives

$$c^T x \leq b^T y^* = c^T x^* \quad \forall x \in D$$

and

$$b^T y \geq c^T x^* = b^T y^* \quad \forall y \in \tilde{D},$$

hence $x^* \in D^*$ and $y^* \in \tilde{D}^*$, as desired.

Next, if the feasible set $\tilde{D}$ is nonempty, then the cost function in (4.1) can not be bounded from above due to (4.5). Similarly, if the feasible set D is nonempty, then the cost function in (4.3) can not be bounded from below due to (4.5). Therefore, the assertions are true. □

Figure 17.1:

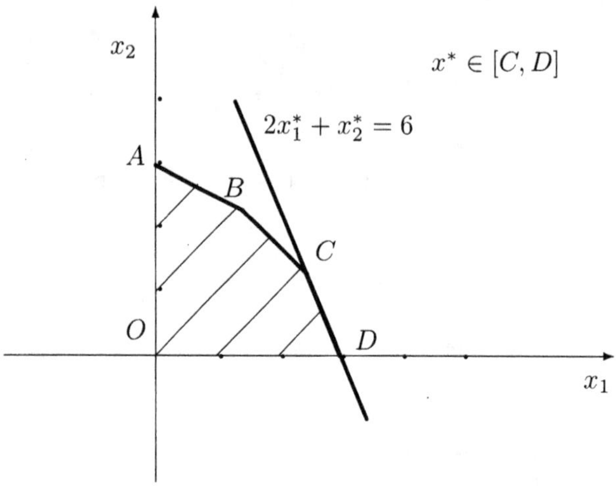

Exercise 4.2 *Find a solution to the following linear programming problem:*

$$\max \rightarrow 2x_1 + x_2$$

subject to

$$3x_1 + 7x_2 \leq 21,$$
$$6x_1 + 3x_2 \leq 18,$$
$$8x_1 + 7x_2 \leq 28,$$
$$x_1 \geq 0, x_2 \geq 0;$$

and a solution to its dual.

The primal problem can be solved by the graphical method (see Figure 17.1). The feasible set is represented by the polygon OABCD. The maximal value of the cost function is attained at the each point of the segment $[C, D]$, where $C = (7/3, 4/3)^T$, $D = (3, 0)^T$ and results in 6. Let us consider the dual problem:

$$\min \rightarrow 21y_1 + 18y_2 + 28y_3$$

subject to

$$3y_1 + 6y_2 + 8y_3 \geq 2,$$

$$7y_1 + 3y_2 + 7y_2 \geq 1,$$
$$y_1 \geq 0, y_2 \geq 0, y_3 \geq 0.$$

Since $3x_1^* + 7x_2^* < 21$, (4.8) yields $y_1^* = 0$ and

$$\begin{cases} 6y_2^* + 8y_3^* = 2, \\ 3y_2^* + 7y_3^* = 1. \end{cases}$$

This system has the unique solution $y_2^* = 1/3$, $y_3^* = 0$. Since y^* is nonnegative it has to be a solution to the dual problem. In fact,

$$21y_1^* + 18y_2^* + 28y_3^* = 6$$

and (4.7) holds. □

Chapter 5

Exercise 5.1 *By using the results of Section 11.2, find existence and uniqueness conditions for the Cassel-Wald model.*

The problem is to find $x^* \in D$ such that

$$(x^* - x)^T c(x^*) \geq 0 \quad \forall x \in D,$$

where

$$D = \{x \in \mathbb{R}^n \mid Ax \leq b, x \geq 0\}.$$

Since A is the consumption rates matrix and b is the endowments vector, they contains only non-negative entries. Then D is clearly nonempty, convex and closed. Moreover, it is natural to suppose that production of each commodity requires some positive amount of at least one resource, i.e. the matrix A has no zero column. Then D is bounded (see Exercise 12.1), hence compact. If the price mapping c is continuous on $\mathbb{R}^n_+$, then, by Theorem 11.2, the above problem has a solution, i.e. there exists at least one equilibrium point. Moreover if we suppose additionally that $-c$ is strictly monotone, then, by Proposition 11.14, this solution is unique. □

Exercise 5.2 *Derive the above optimality conditions for problem (5.4).*

Setting $\varphi(\alpha) = f_j(\alpha) - \alpha p^T a^j$, we can rewrite (5.4) as follows:

$$\min_{\alpha \geq 0} \to \varphi(\alpha),$$

so that $\varphi'(\alpha) = c_j(\alpha) - p^T a^j$. Combining Theorem 11.1 and Proposition 7.1, we obtain the optimality conditions at x_j:

$$x_j \geq 0, \varphi'(x_j) \geq 0, x_j\varphi'(x_j) = 0;$$

and the result follows. □

Exercise 5.3 *Show that the transportation problems are particular cases of* (4.1′), (4.2′) *and* (4.3′), (4.4′).

We note that we have two linear programming problems:

$$\begin{array}{ll} \max \to \sum\limits_{i=1}^{l} d_i p_i & \min \to \sum\limits_{i=1}^{l}\sum\limits_{j=1}^{l} c_{ij} f_{ij} \\ p_j - p_i \leq c_{ij} & \sum\limits_{j=1}^{l} f_{ji} - \sum\limits_{j=1}^{l} f_{ij} \geq d_i \\ i,j = 1,\ldots,l; & i = 1,\ldots,l; \\ p_i \geq 0 \quad i = 1,\ldots,l; & f_{ij} \geq 0 \quad i,j = 1,\ldots,l; \end{array}$$

since all the cost and constraint functions are affine. Moreover, the basic relationships between the numbers of variables and constraints, and between the coefficients of the cost functions and right-hand sides of constraints are fulfilled. Also, their basic constraints have opposite signs. The constraint matrix A has the dimensionality $l^2 \times l$, where the (i,j)-th row contains two nonzero components $a_{(ij),j} = 1$ and $a_{(ij),i} = -1$. But the transposed matrix A^T then determines the constraint coefficients of the second problem. Therefore, the problems are mutually dual. □

Exercise 5.4 *By using the results of Section 11.2, find existence and uniqueness conditions for the model with fixed capacities.*

Let us consider the variational inequality: Find $f^* \in D$ such that

$$\sum_{i=1}^{l}\sum_{j=1}^{l} c_{ij}(f^*)(f_{ij} - f^*_{ij}) \geq 0 \quad \forall f \in D, \tag{17.1}$$

where

$$D = \left\{ f \in \mathbb{R}_+^{l\times l} \,\middle|\, \sum_{j=1}^{l} f_{ji} - \sum_{j=1}^{l} f_{ij} \geq d_i \quad i = 1,\ldots,l \right\}.$$

If VI (17.1) has a solution f^*, then, applying Proposition 11.6 (ii), we see that there exists a point $p^* \in \mathbb{R}^l_+$ such that

$$\sum_{i=1}^{l}\sum_{j=1}^{l}\left(c_{ij}(f^*) - p_j^* + p_i^*\right)(f_{ij} - f_{ij}^*) \geq 0 \quad \forall f \geq 0,$$

$$\sum_{i=1}^{l}\left[\sum_{j=1}^{l} f_{ji}^* - \sum_{j=1}^{l} f_{ij}^* - d_i\right](p_i - p_i^*) \geq 0 \quad \forall p \geq 0;$$

i.e. the pair (f^*, p^*) is an equilibrium point for the model with fixed capacities. Let us turn to VI (17.1). The feasible set D is convex and closed. It is natural to suppose that D is nonempty. Due to Proposition 11.14, if c is strictly monotone, then VI (17.1) has at most one solution. Next, due to Theorem 11.4, if c is continuous and strongly monotone, then VI (17.1) has a unique solution. □

Exercise 5.5 *Let the functions p and h_i be affine, i.e.,*

$$\begin{aligned} p(\sigma) &= \alpha - \beta\sigma, \alpha \geq 0, \beta > 0; \\ h_i(x_i) &= \gamma_i x_i + \delta_i, \gamma_i \geq 0, \delta_i \geq 0 \quad \textit{for} \quad i = 1, \dots, n. \end{aligned}$$

Then,

$$f_i(x) = x_i(\alpha - \beta\sigma_x) - \gamma_i x_i - \delta_i.$$

Find the value of $G(x)$ and show that the oligopolistic equilibrium model becomes equivalent to the problem of minimizing a strongly convex quadratic function subject to the nonnegativity constraints.

By definition,

$$\begin{aligned} G_i(x) &= -p(\sigma_x) - x_i p'(\sigma_x) + h_i'(x_i) \\ &= -\alpha + \beta\sigma_x + \beta x_i + \gamma_i \end{aligned}$$

for $i = 1, \dots, n$, where $\sigma_x = \sum_{j=1}^{n} x_j$. Hence,

$$G(x) = \beta(Ax - b),$$

where

$$A = \begin{pmatrix} 2 & 1 & \dots & 1 \\ 1 & 2 & \dots & 1 \\ \vdots & \vdots & \ddots & \vdots \\ 1 & 1 & \dots & 2 \end{pmatrix},$$

$b = (b_1, \ldots, b_n)^T$, $b_i = (\alpha - \gamma_i)/\beta$ for $i = 1, \ldots, n$. From Corollary 11.2 it follows that the oligopolistic equilibrium problem is equivalent to the following VI: Find $x^* \geq 0$ such that

$$(x - x^*)^T (Ax^* - b) \geq 0 \quad \forall x \geq 0,$$

however, $Ax - b = \nabla\varphi(x)$, where $\varphi(x) = 0.5x^T Ax - x^T b$ since A is symmetric. Moreover, A is positive definite and φ must be strongly convex. Due to Theorem 11.1, the above VI is equivalent to the optimization problem

$$\min_{x \geq 0} \to \varphi(x).$$

On account of Proposition 11.15 (ii) it has a unique solution. □

Exercise 5.6 *Set* $n = 2, q_i(p) = \alpha p_{3-i}/p_i - \beta, h_i(t) = \gamma t + \delta$ *with* $\alpha, \beta, \gamma, \delta > 0$ *for* $i = 1, 2$ *and find the equilibrium point via solution of system (5.23).*

By definition,

$$f_i(p) = \alpha p_{3-i} - \beta p_i - \gamma\left(\alpha p_{3-i}/p_i - \beta\right) - \delta$$

and

$$G_i(p) = \beta - \alpha\gamma p_{3-i}/p_i^2 \quad \text{for} \quad i = 1, 2.$$

System (5.23) is specialized as follows:

$$\begin{cases} \beta p_1^2 - \alpha\gamma p_2 = 0, \\ \beta p_2^2 - \alpha\gamma p_1 = 0. \end{cases}$$

It has the unique solution $p_1 = p_2 = \alpha\gamma/\beta$. □

Exercise 5.7 *By using the theorems of Section 11.2, find existence and uniqueness results for problem (5.25).*

The problem is to find $x^* \in X$ such that

$$(x - x^*)^T p(x^*) \geq 0 \quad \forall x \in X,$$

where

$$X = \left\{ x \in \mathbb{R}^n_+ \;\middle|\; \sum_{i=1}^n x_i = b \right\}.$$

Clearly, X is a nonempty, convex, and compact set. If the mapping p is continuous on X, then, by Theorem 11.2, the problem is solvable. Moreover, if we suppose additionally that p is strictly monotone, then, by Proposition 11.14, its solution is unique. □

Exercise 5.8 *Deduce the equivalence of (5.26) and (5.27) from Proposition 11.7. By using the theorems of Section 11.2, find existence and uniqueness results for problem (5.27).*

From Proposition 11.7 we obtain the necessary and sufficient optimality conditions for VI (5.27): $(x^*, \lambda) \in [\alpha, \beta] \times \mathbb{R}$ and

$$\sum_{i=1}^{n} (p_i(x^*) - \lambda)(x_i - x_i^*) \geq 0 \quad \forall x_i \in [\alpha_i, \beta_i], \quad i = 1, \ldots, n;$$

$$\left(\sum_{i=1}^{n} x_i^* - b \right) (\mu - \lambda) \geq 0 \quad \forall \mu \in \mathbb{R};$$

where $\alpha = (\alpha_1, \ldots, \alpha_n)^T$, $\beta = (\beta_1, \ldots, \beta_n)^T$. The first inequality is equivalent to (5.26); see Proposition 7.2. At the same time, the second relation is rewritten as $\sum_{i=1}^{n} x_i^* = b$, i.e. $x^* \in \tilde{X}$.

Next, $\tilde{X}$ is convex and compact. If $\tilde{X}$ is nonempty and p is continuous on $\tilde{X}$, then, by Theorem 11.2, VI (5.27) is solvable. If we suppose additionally that p is strictly monotone, then, by Proposition 11.14, (5.27) has a unique solution. □

Exercise 5.9 *Find solutions of problem (5.26) when $n = 2$, $\alpha_1 = 1$, $\beta_1 = 2$, $\alpha_2 = 1$, $\beta_2 = 3$, $p_1(x_1) = 2x_1 - 1$, $p_2(x_2) = 0.5x_2 + 1$ for $b = 3$ and for $b = 5$ and give their graphical illustration.*

We first consider the reduced problem

$$\begin{cases} p_i(x_i^*) = \lambda & i = 1, 2, \\ x_1^* + x_2^* = b \end{cases}$$

and then check whether its solution satisfies the offer bounds.

In case $b = 3$ we have the system

$$\begin{cases} 2x_1^* - 1 = \lambda, \\ 0.5x_2^* + 1 = \lambda, \\ x_1^* + x_2^* = 3; \end{cases}$$

which has the unique solution $x_1^* = 1.4, x_2^* = 1.6, \lambda = 1.8$. This point satisfies the bounds and solves also the initial problem. Its graphical interpretation is given in Figure 17.2.

Figure 17.2:

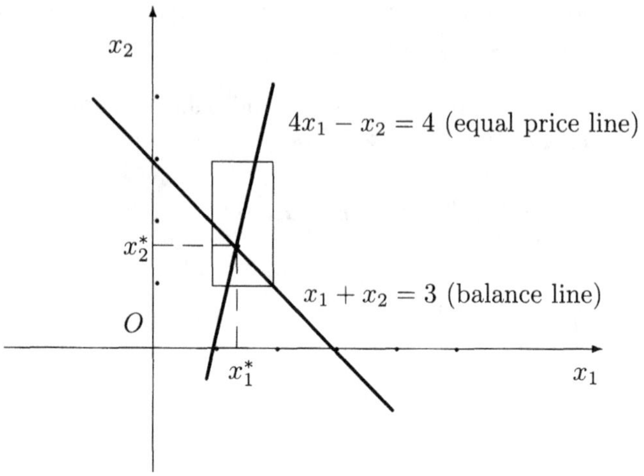

In case $b = 5$ we have the system

$$\begin{cases} 2\tilde{x}_1 - 1 = \lambda, \\ 0.5\tilde{x}_2 + 1 = \lambda, \\ \tilde{x}_1 + \tilde{x}_2 = 5; \end{cases}$$

which has the unique solution $\tilde{x}_1 = 1.8$, $\tilde{x}_2 = 3.2$, $\lambda = 2.6$, but it is infeasible for the second trader. We reduce its offer by setting $x_2^* = 3$, then $x_1^* = 2$. Observe that $p_1(x_1^*) = 3$ and $p_2(x_2^*) = 2.5$, hence we set $\lambda = 3$ and obtain $p_i(x_i^*) \leq \lambda$, $x_i^* = \beta_i$ for $i = 1, 2$, and (5.26) is fulfilled. Figure 17.3 gives the graphical illustration. □

Exercise 5.10 *By using the theorems of Section 11.2, find the general existence result for problem (5.28), (5.29).*

The feasible set Z, defined by (5.28), is nonempty, convex, and closed. Since $0 \in Z$, we specialize Theorem 11.3 by setting $y = 0$ there. Suppose that the price mappings g_i and h_j are continuous and that there exists a bounded set $Y \subseteq Z$ such that $0 \in Y$ and

$$\sum_{i=1}^{n} g_i(x, y)x_i - \sum_{j=1}^{m} h_j(x, y)y_j > 0 \quad \text{for every } (x, y) \in Z \setminus Y.$$

Figure 17.3:

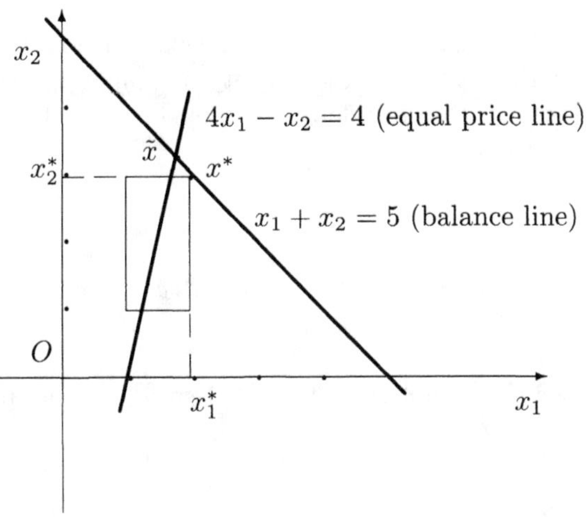

Then problem (5.28), (5.29) has a solution. □

Chapter 6

Exercise 6.1 *By using the results of Section 11.2, find existence conditions for the path flow formulation of the network equilibrium model.*

The feasible set X, defined by (6.4), is nonempty, convex, and compact. If we suppose that the cost mappings T_l in (6.5) are continuous, then, by Theorem 11.2, problem (6.4)–(6.6), (6.8) has a solution. □

Exercise 6.2 *Prove that H in (6.9) is bounded.*

Due to the constraints

$$\mathbf{h} \geq 0, \ \sum_{j \neq i} h_{ij} \leq b_i,$$

the feasible flows $\mathbf{h}$ are bounded. Hence the population volumes

$$x_i = b_i + \sum_{j \neq i} h_{ji} - \sum_{j \neq i} h_{ij}, i \in \mathbf{N},$$

have to be bounded and the result follows. □

Exercise 6.3 *By using the results of Section 11.2, find existence conditions for the migration equilibrium model.*

By definition, the set H in (6.9) is nonempty, convex, and closed, moreover, it is bounded due to Exercise 6.2. If the utility mappings u_i and the migration cost mappings c_{ij} are continuous, then, by Theorem 11.2, problem (6.9), (6.12) has a solution. □

Chapter 7

Exercise 7.1 *Prove that problem (7.3) is equivalent to CP (7.1) with* $X = \mathbb{R}^n_+$.

Setting $X = \mathbb{R}^n_+$ in (7.1) and noticing that $(\mathbb{R}^n_+)' = \mathbb{R}^n_+$, we obtain

$$x^* \geq 0, G(x^*) \geq 0, (x^*)^T G(x^*) = 0.$$

Since $x_i^* G_i(x^*) \geq 0$, we see that the third equality is equivalent to the sequence of equalities: $x_i^* G_i(x^*) = 0$ for $i = 1, \ldots, n$. Therefore, the above problem coincides with CP (7.3). □

Exercise 7.2 *Prove the assertions of Proposition 7.2.*

Obviously, for each $i = 1, \ldots, n$, the inequality

$$G_i(x^*)(x_i - x_i^*) \geq 0 \quad \forall x_i \in [\alpha_i, \beta_i]$$

is rewritten equivalently as follows:

$$G_i(x^*) \begin{cases} \geq 0 & \text{if} \quad x_i^* = \alpha_i, \\ = 0 & \text{if} \quad x_i^* \in (\alpha_i, \beta_i), \\ \leq 0 & \text{if} \quad x_i^* = \beta_i; \end{cases}$$

see also Exercise 1.1. Hence (i) and (ii) are equivalent. Next, if $x^* \in X$ satisfies the sequence of inequalities

$$G_i(x^*)(x_i - x_i^*) \geq 0 \quad \forall x_i \in [\alpha_i, \beta_i], \ \forall i = 1, \ldots, n;$$

then their summing gives that x^* solves VI (7.2), (7.4). Conversely, if x^* solves VI (7.2), (7.4), for each k, we can take the point

$$\tilde{x} = (x_1^*, \ldots, x_{k-1}^*, x_k, x_{k+1}^*, \ldots, x_n^*) \in X$$

with any $x_k \in [\alpha_k, \beta_k]$ and, setting $x = \tilde{x}$ in (7.2), we obtain

$$G_k(x^*)(x_k - x_k^*) \geq 0,$$

hence x^* solves the above sequence of inequalities, and the result follows. □

Exercise 7.4 *Following the proof of Proposition 7.3, establish the result of Proposition 7.4.*

Clearly, (i) ⇒ (ii) for $n = 1$. Using induction, we suppose that this implication holds for $n-1$ with $n > 1$ and that there exist a P_0-matrix A and a nonzero vector $z \in \mathbb{R}^n$ such that

$$\sum_{j=1}^{n} a_{kj} z_k z_j < 0 \quad \text{for each } \ k \ \text{ with } z_k \neq 0.$$

If $z_i = 0$ for some i, then there exists a principal submatrix $\tilde{A}$ of A, which is also a P_0-matrix such that

$$\sum_{j \in J} \tilde{a}_{kj} \tilde{z}_k \tilde{z}_j < 0 \quad \text{for each } \ k \text{ with } \tilde{z}_k \neq 0.$$

and for a subvector $\tilde{z}$ of z, where J is the corresponding subset of indices. Since this is a contradiction, no component of z is zero. Set

$$d_i = \frac{1}{z_i} \sum_{j=1}^{n} a_{ij} z_j < 0 \quad \text{for} \quad i = 1, \dots, n$$

and define the diagonal matrix D with the diagonal entries $d_1, \dots, d_n$. It follows that $(A - D)z = 0$, but the matrix $A - D$ has to be nonsingular, a contradiction, Hence, (i) ⇒ (ii).

Suppose that (ii) holds and choose an arbitrary real eigenvalue λ of A and an associate eigenvector z. The vector z must be nonzero and, as λ is real, we can take z to be real. By definition, $Az = \lambda z$ and there exists an index k such that $z_k \neq 0$ and

$$0 \leq \sum_{j=1}^{n} a_{kj} z_k z_j = \lambda z_k^2,$$

hence $\lambda \geq 0$ and (ii)⇒ (iii).

Next, let (iii) hold. The determinant of A is equal to the product of all its eigenvalues, but the complex eigenvalues of real matrices appear in conjugate pairs. It follows that A is a P_0-matrix, i.e. (i) holds, and the proof is complete. □

Exercise 7.6 *Derive from Proposition 7.6 the feasibility of LCP (7.3) with $G(x) = Ax + b$, where A is a P-matrix.*

By Proposition 7.6, there exists $x > 0$ such that $Ax > 0$. For arbitrary b, there exists $\lambda > 0$ such that $A(\lambda x) + b \geq 0$. Hence, the point $\tilde{x} = \lambda x$ is feasible. □

Exercise 7.7 *Define*

$$D = \{x \in D \mid x_i < \beta_i \Rightarrow G_i(x) \geq 0 \quad \forall i = 1, \ldots, n\}$$

for MCP (7.2), (7.4) and show that the assertion of Theorem 7.2 remains true if $\alpha_i > -\infty$ for $i = 1, \ldots, n$ and $\mathbb{R}^n_+$ is replaced with X. Make the modification of the Jacobi algorithm which allows for obtaining the result of Theorem 7.3.

Since D is then bounded below, we should follow the proofs of Lemmas 7.1, 7.2 and Theorem 7.2 in general. In the modified Jacobi algorithm, we also compute the next iterate componentwise. For each index $i = 1, \ldots, n$, we set $x_i^{k+1} = \alpha_i$ if $G_i(x_{-i}^k, \alpha_i) \geq 0$ and $x_i^{k+1} = \beta_i$ if $x_i^k = \beta_i$ and $G_i(x_{-i}^k, \beta_i) \leq 0$. Otherwise we set x_i^{k+1} to be the number in $(\alpha_i, x_i^k]$ such that $G_i(x_{-i}^k, x_i^{k+1}) = 0$. □

Chapter 8

Exercise 8.1 *Replace D in (8.4) by*

$$D = \left\{ x \in X \;\middle|\; x_i < x_i'' \Longrightarrow x_i - \sum_{j=1}^{n} a_{ij} x_j - y_i \geq 0 \textit{ for } \quad i = 1, \ldots, n \right\}$$

and show that the assertion of Theorem 8.1 remains true for problem (8.3), (8.5).

Since the cost mapping $x \mapsto (I - A)x - y$ is Z, follow the lines of Exercise 7.7. □

Exercise 8.2 *Describe the modification of the Jacobi algorithm for MCP (8.3), (8.5) and prove its convergence.*

Follow the lines of the solution to Exercise 7.7. □

Exercise 8.3 *Prove the assertions of Theorem 8.3.*

The result follows directly from Proposition 8.1 and Theorems 7.2 and 7.3. □

Exercise 8.5 *Prove formula (8.10).*

Observe that x^i is the unique solution of the optimization problem:

$$\max \to \sum_{j=1}^{n} \sigma_{ij} \ln x_j$$

subject to

$$p^T x \leq p^T w^{(i)}, x \geq 0.$$

Clearly, if we set

$$x^i_j = \frac{1}{p_j} \sigma_{ij} p^T w^{(i)} \quad \text{for} \quad j = 1, \ldots, n,$$

then x^i is non-negative, moreover, the optimality conditions

$$\frac{\sigma_{ij}}{x^i_j} - \lambda p_j = 0 \quad \text{for} \quad j = 1, \ldots, n; \quad p^T x^i = p^T w^{(i)}$$

hold true with $\lambda = 1/p^T w^{(i)}$; see Proposition 11.7. Hence x^i in (8.10) solves the optimization problem. □

Exercise 8.6 *Show that f_i is concave in p_i under the above assumptions.*

By definition,

$$\begin{aligned} & f_i(p_1, \ldots, p_n) \\ & = \alpha_i p_i \left[\eta(p_1, \ldots, p_{i-1}, p_{i+1}, \ldots, p_n)/p_i\right]^\kappa - \beta_i p_i \\ & -\alpha_i \gamma_i \left[\eta(p_1, \ldots, p_{i-1}, p_{i+1}, \ldots, p_n)/p_i\right]^\kappa + \gamma_i \beta_i + \delta_i, \end{aligned}$$

where $\alpha_i > 0$, $\beta_i > 0$, $\kappa \in (0, 1]$, $\eta : \mathbb{R}^{n-1} \to \mathbb{R}_+$ is a non-negative differentiable function, which is non-decreasing in each variable. It follows that the partial derivative

$$\frac{\partial f_i(p)}{\partial p_i} = \kappa \alpha_i \gamma_i p_i^{-1} \left[\eta(p_{-i})/p_i\right]^\kappa + (1 - \kappa) \alpha_i \left[\eta(p_{-i})/p_i\right]^\kappa - \beta_i$$

with $p_{-i} = (p_1, \ldots, p_{i-1}, p_{i+1}, \ldots, p_n)$ is non-increasing in p_i, hence f_i is concave in p_i. □

Chapter 9

Exercise 9.3 *Let X be a set of form (7.4). Prove that $G : X \to \mathbb{R}^n$ is a P_0-(respectively, P-) mapping, if it is monotone (respectively, strictly monotone), but the reverse assertions are not true.*

Clearly, if the sum

$$\sum_{i=1}^{n} (x_i' - x_i'')[G_i(x') - G_i(x'')]$$

is non-negative (positive), then so is the expression

$$(x_i' - x_i'')[G_i(x') - G_i(x'')]$$

for at least one i. On the second part, see Exercise 7.3 and Example 7.1.□

Exercise 9.4 *Suppose that G is of form (9.1). Prove that G is monotone (respectively, strictly monotone) if and only if A is positive semidefinite (respectively, positive definite).*

If $G(x) = Ax + b$, then

$$(x' - x'')^T [G(x') - G(x'')] = (x' - x'')^T A(x' - x'')$$

and the result follows from the definitions. □

Exercise 9.5 *Prove the extensions of Theorems 9.1 and 9.2 for MCP (7.2), (7.4).*

The assertion of Theorem 9.1 remains true for the general MCP (7.2), (7.4) with small modifications of the proof. Now we turn to Theorem 9.2 and consider MCP (7.2), (7.4) under the following boundedness condition:

$$\alpha_i > -\infty \quad \text{for } i = 1, \ldots, n.$$

Next, set

$$J = \{i \mid \beta_i = +\infty\}.$$

Theorem 17.1. *(i)* If $G : X \to E$ is a continuous and monotone mapping and there exists a point $y \in X$ such that $G_i(y) > 0$ for $i \in J$, then MCP (7.2), (7.4) has a solution.

(ii) If $G : X \to \mathbb{R}^n$ is a continuous and strictly monotone mapping and there exists a point $y \in X$ such that $G_i(y) \geq 0$ for $i \in J$, then MCP (7.2), (7.4) has a unique solution.

Proof. The case $J = \emptyset$ is trivial since the results follow from Theorem 11.2 and Proposition 11.14. Let $J \neq \emptyset$. Take an arbitrary $x \in X$. Then, by monotonicity,

$$(x - y)^T G(x) \geq x^T G(y) - y^T G(y).$$

Since $G_i(y) > 0$ for all $i \in J$, there exists a number $\gamma > 0$ such that the right-hand side of the above inequality is positive, if $x \in X \backslash Y$, where

$$Y = \left\{ x \in X \,\middle|\, \sum_{i=1}^{n} x_i \leq \gamma \right\}.$$

Assertion (i) follows now from Theorem 11.3.

In case (ii), it suffices to prove that the conditions of part (i) are fulfilled. Fix a number $\gamma > 0$ so that the set

$$K = \left\{ z \in X \,\middle|\, z \geq y, \sum_{i=1}^{n} z_i = \sum_{i=1}^{n} y_i + \gamma \right\}$$

is nonempty. Since K is convex and compact, Theorem 11.2 with $X = K$ guarantees the existence of a solution $\tilde{x} \in K$ for the following VI:

$$(z - \tilde{x})^T [G(\tilde{x}) - G(y)] \geq 0 \quad \forall z \in K.$$

Since $\tilde{x} \neq y$, we have

$$(\tilde{x} - y)^T [G(\tilde{x}) - G(y)] > 0,$$

therefore,

$$(z - y)^T [G(\tilde{x}) - G(y)] > 0 \quad \forall z \in K.$$

If we take $z \in K$ such that $z_j = y_j$ for $j \notin J$, the above inequality gives $G_i(\tilde{x}) > 0$ for all $i \in J$ and the result follows. □

Exercise 9.6 *By using the projection properties, prove the formulas for computation of* y_i^α *and* $y^\alpha(x)$.

By definition, $y^\alpha(x)$ solves the optimization problem

$$\min_{y \in \mathbb{R}^n_+} \to (2\alpha)^{-1}\|x-y\|^2 - (x-y)^T G(x),$$

which can be replaced by VI (see Theorem 11.1):

$$(z - y^\alpha(x))^T(y^\alpha(x) - [x - \alpha G(x)]) \geq 0 \quad \forall z \in \mathbb{R}^n_+.$$

Due to Proposition 11.12, this means that $y^\alpha(x) = \pi_+[x - \alpha G(x)]$. On account of (9.3), it can be computed componentwise, i.e. $y_i^\alpha = \max\{0, x_i - \alpha G_i(x)\}$. □

Chapter 10

Exercise 10.2 *Prove the assertion of Theorem 10.2.*

Utilize Corollary 11.2, Lemma 10.1 and Exercise 11.2. □

Exercise 10.3 *Extend Theorems 10.2–10.5 for the case of bounded outputs.*

We should simply replace CP (9.2) by MCP (7.2), (10.9), with the bounded feasible set

$$X = [0, \beta_1] \times \cdots \times [0, \beta_n].$$

Of course, the extension of Theorem 10.2 is trivial. Next, by Theorem 11.2, this MCP is solvable and the conditions of Theorem 10.3 and 10.4 yield also the uniqueness of this solution. □

Exercise 10.4 *Give the sufficient conditions of convergence of the algorithm of Section 9.2 for CP (9.2), (10.9).*

Under the assumptions of either Proposition 10.4 (iii) or Proposition 10.5 (i), $\nabla G(x)$ is a P-matrix. If we suppose also that φ_α has bounded level sets, then the algorithm ensures the convergence to a solution. □

Chapter 11

Exercise 11.1 *Prove the assertion of Proposition 11.7.*

Replacing each i-th equality $h_i(u) = 0$ with two inequalities $h_i(u) \leq 0$ and $h_i(u) \geq 0$ and applying Proposition 11.6, we see that the optimality conditions are written as follows: Find $(u^*, v^*, p^*, q^*) \in U \times \mathbb{R}^m_+ \times \mathbb{R}^k_+ \times \mathbb{R}^k_+$ such that

$$(u-u^*)^T \left[F(u^*) + \sum_{i=1}^{m} v_i^* a^i + \sum_{j=1}^{k} p_j^* b^j - \sum_{j=1}^{k} q_j^* b^j \right] \geq 0$$

$$\forall u \in U,$$
$$(v-v^*)^T[-f(u^*)] \geq 0 \quad \forall v \in \mathbb{R}^m_+,$$
$$(p-p^*)^T[-h(u^*)] \geq 0 \quad \forall p \in \mathbb{R}^k_+,$$
$$(q-q^*)^T[h(u^*)] \geq 0 \quad \forall q \in \mathbb{R}^k_+.$$

Of course, it we set $w = p - q \in \mathbb{R}^k$ and $w^* = p^* - q^* \in \mathbb{R}^k$, then the last two inequalities become equivalent to the following

$$(w-w^*)^T[-h(u^*)] \geq 0 \quad \forall w \in \mathbb{R}^k,$$

and the system coincides with (11.1), (11.19). □

Exercise 11.2 *Prove the equivalence of problems (11.21)-(11.23) and (11.20).*

Clearly, summation of the inequalities in (11.20) gives (11.23). Conversely, if $x^* = (u_1^*, \ldots, u_m^*)$ solves (11.21)–(11.23), then setting

$$y = (u_1^*, \ldots, u_{i-1}^*, v_i, u_{i+1}^*, \ldots, u_m^*)$$

in (11.23) for any $v_i \in X_i$ and for an arbitrary index i gives (11.20). □

Exercise 11.4 *Show that the saddle point problem (11.7), (11.29) is equivalent to the system (11.26), (11.11) with $F(u) = \nabla f_0(u)$, $b(v) = \nabla\varphi(v)$ under the above assumptions.*

Apply Corollary 11.1. □

Chapter 12

Exercise 12.1 *Prove that D is bounded if A and b contain only non-negative entries and there is no zero column in A.*

By definition,

$$D = \{x \in \mathbb{R}^n \mid Ax \leq b, x \geq 0\},$$

and for each j there is i such that $a_{ij} > 0$. Hence

$$0 \le x_j \le \frac{1}{a_{ij}} \left(b_i - \sum_{k \neq j} a_{ik} x_k \right) \le b_i / a_{ij}$$

and D is bounded. □

Exercise 12.2 *Prove the assertion of Proposition 12.1.*

Since Y is nonempty and compact, problem (12.5) has a solution. Suppose that y' and y'' are two different solutions of (12.5), then $p^T y' = p^T y'' = \mu$ and $\tilde{y} = \alpha y' + (1 - \alpha) y''$ is also a solution of (12.5) if $\alpha \in (0, 1)$. But $\tilde{y} \in \text{int} Y$ and there exists $\delta > 0$ such that $\tilde{y} + \delta p / \|p\| \in Y$. Then we obtain a contradiction, since

$$p^T(\tilde{y} + \delta p / \|p\|) = p^T \tilde{y} + \delta \|p\|^2 = \mu + \delta \|p\|^2 > \mu.$$

□

Exercise 12.3 *Prove the assertion of Proposition 12.2.*

Clearly, $y(\lambda p) = y(p)$ for every $\lambda > 0$, hence the mapping $p \mapsto y(p)$ is positively homogeneous of degree 0. Take any $p', p'' \in \mathbb{R}^n_+ \setminus \{0\}$ and set $y' = y(p'), y'' = y(p'')$. Then we have

$$(p')^T(y' - y'') \ge 0 \quad \text{and} \quad (p'')^T(y'' - y') \ge 0,$$

hence

$$(p' - p'')^T(y' - y'') \ge 0$$

and the mapping is monotone. Next, take a sequence $\{p^k\} \to \tilde{p} \in \mathbb{R}^n_+ \setminus \{0\}$ and set $y^k = y(p^k)$. Then $y^k \in Y$ and

$$(p^k)^T(y^k - y) \ge 0 \quad \forall y \in Y.$$

Since $\{y^k\}$ is bounded, we can take the limit $k \to \infty$ and a subsequence, if necessary, and obtain

$$(\tilde{p})^T(\tilde{y} - y) \ge 0 \quad \forall y \in Y,$$

where $\tilde{y}$ is a limit point of $\{y^k\}$. It follows that $\tilde{y} \in Y$ and $\tilde{y} = y(\tilde{p})$. Since $y(\tilde{p})$ is determined uniquely, we have $\tilde{y} = \lim_{k \to \infty} y^k$ and the result follows. □

Exercise 12.4 *Prove that the mapping* $p \mapsto x(p)$ *is continuous on* $\mathbb{R}^n_+ \setminus \{0\}$.

Utilize Lemma 12.1 and follow the lines of the above proof. □

Exercise 12.5 *Prove the assertion of Proposition 12.4.*

Since each cost function h_i is convex and each function

$$\theta_i(x, y_i) = -y_i p\left(\sum_{j \neq i} x_j + y_i\right)$$

is concave in x, it suffices to prove that the function $\varphi(x) = \sigma_x p(\sigma_x)$ is concave. Since the industry revenue function $\mu(\sigma) = \sigma p(\sigma)$ is concave, we see that, for all $x', x'' \in \mathbb{R}^n_+$ and for each $\lambda \in (0,1)$, it holds that

$$\begin{aligned} \varphi(\lambda x' + (1-\lambda)x'') &= \mu(\lambda\sigma_{x'} + (1-\lambda)\sigma_{x''}) \\ &\geq \lambda\mu(\sigma_{x'}) + (1-\lambda)\mu(\sigma_{x''}) \\ &= \lambda\varphi(x') + (1-\lambda)\varphi(x''), \end{aligned}$$

i.e. φ is concave and the result follows. □

Exercise 12.6 *Establish monotonicity criteria for the cost mapping and find existence and uniqueness results for VI (5.28), (5.29).*

Being based on Proposition 11.1, we can obtain directly the monotonicity properties of the mapping

$$(x, y) \mapsto (g(x, y), -h(x, y))$$

via consideration of positive (semi) definiteness of its Jacobian. The general existence result for continuous mappings g and h is given in Exercise 5.10. If the cost mapping is strictly monotone, then, by Proposition 11.14, VI (5.28), (5.29) has at most one solution. If the cost mapping is continuous and strongly monotone, then, by Theorem 11.4, the problem has a unique solution. □

Exercise 12.7 *Prove that the monotonicity of* F *implies the monotonicity of* G.

By definition,

$$G(x) = B^T F(Bx).$$

Choose x', x'' and set $y' = Bx', y'' = Bx''$. Then

$$\begin{aligned}&(x' - x'')^T [G(x') - G(x'')]\\&= (x' - x'')^T \left[B^T F(y') - B^T F(y'')\right]\\&= (y' - y'')^T [F(y') - F(y'')] \geq 0,\end{aligned}$$

i.e. G is monotone. □

Chapter 13

Exercise 13.2 *Let $G(x) = Ax + b$, where A is a symmetric positive semidefinite $n \times n$ matrix, $b \in \mathbb{R}^n$. Prove that G is co-coercive.*

In this case, there exists a matrix Q such that $A = Q^T Q$ and $\|Q^T\|^2 = \|A\|$. Then, for arbitrary $x, y \in \mathbb{R}^n$, we have

$$\begin{aligned}&(x - y)^T [G(x) - G(y)]\\&= (x - y)^T A(x - y) = \|Q(x - y)\|^2\\&\geq \|Q^T\|^{-2}\|Q^T Q(x - y)\|^2 = \frac{1}{\|A\|}\|A(x - y)\|^2\\&= \frac{1}{\|A\|}\|G(x) - G(y)\|^2,\end{aligned}$$

and the result follows. □

Exercise 13.3 *Prove that assumption (13.6) is not fulfilled if $G(x) = Ax + b$, where A is an arbitrary skew-symmetric matrix.*

If $x^* \in X^*$, then for every $x \in X$, we have

$$\begin{aligned}&(x - x^*)^T G(x)\\&= (x - x^*)^T [G(x) - G(x^*)] + (x - x^*)^T G(x^*)\\&= (x - x^*)^T A(x - x^*) + (x - x^*)^T G(x^*)\\&= (x - x^*)^T G(x^*).\end{aligned}$$

Hence (13.6) does not hold if there is a point $x \in X$ such that $(x - x^*)^T G(x^*) = 0$. □

Chapter 14

Exercise 14.2 *Prove that both (14.2) and (14.3) imply (13.9), i.e. the integrability of G.*

In case (14.2) we have

$$f(x) = \sum_{i=1}^{n} f_i(x_i),$$

where $f_i(x_i) = \int_0^{x_i} G_i(\tau)d\tau$ for $i = 1, \ldots, n$. In case (14.3) we have

$$f(x) = 0.5x^T Ax + x^T b.$$

□

Exercise 14.4 *Prove that the integrability of F implies the integrability of G in (14.4).*

If $F(y) = \nabla\varphi(y)$, then, by using the formula for the derivative of a composite function, we have $G(x) = \nabla f(x)$, where $f(x) = \varphi(Bx)$. □

Bibliography

[1] B. Ahn (1979), Computation of Market Equilibrium for Policy Analysis: The Project Independent Evaluation System (PIES), Garland, New York.

[2] K.J. Arrow and F.H. Hahn (1971), General Competitive Analysis, Holden Day, New York.

[3] K.J. Arrow, L. Hurwicz, and H. Uzawa (1958), Studies in Linear and Nonlinear Programming, Stanford University Press, Stanford.

[4] J.-P. Aubin (1998), Optima and Equilibria, Springer-Verlag, Berlin.

[5] A.B. Bakushinskii and A.V. Goncharskii (1989), Iterative Solution Methods for Ill-Posed Problems, Nauka, Moscow; English translation: Ill-Posed Problems: Theory and Applications, Kluwer, Dordrecht, 1994.

[6] K.C. Border (1985), Fixed Point Theorems with Applications to Economics and Game Theory, Cambridge University Press, Cambridge.

[7] R.W. Cottle, J.-S. Pang, and R.E. Stone (1992), The Linear Complementarity Problem, Academic Press, New York.

[8] A. Cournot (1838), Recherches sur les Principles Mathématiques de la Théorie des Richesses, Paris; English translation: Researches into the Mathematical Principles of the Theory of Wealth, N. Bacon, ed., Macmillan, New York, 1897.

[9] G.B. Dantzig (1963), Linear Programming and Extensions, Princeton University Press, Princeton.

[10] I.I. Eremin (1998), Theory of Linear Optimization, Publishing Office of the Ural Branch of RAS, Ekaterinburg; English translation in VSP, Leiden, 2002.

[11] F. Facchinei and J.-S. Pang (2003), Finite-Dimensional Variational Inequalities and Complementarity Problems, Springer-Verlag, Berlin (two volumes).

[12] F. Giannessi and A. Maugeri, Editors (1995), Variational Inequalities and Network Equilibrium Problems, Plenum Press, New York.

[13] E.G. Gol'shtein and N.V. Tret'yakov (1989), Modified Lagrange Functions, Nauka, Moscow; English translation: Modified Lagrangians and Monotone Maps in Optimization, John Wiley and Sons, New York, 1996.

[14] P.T. Harker, Editor (1985), Spatial Price Equilibrium: Advances in Theory, Computation and Application, Springer-Verlag, Berlin.

[15] M. Intriligator (1971), Mathematical Optimization and Economic Theory, Prentice - Hall, New York.

[16] G. Isac (1992), Complementarity Problems, Springer-Verlag, Berlin.

[17] I.V. Konnov (2001), Combined Relaxation Methods for Variational Inequalities, Springer-Verlag, Berlin.

[18] K. Lancaster (1968), Mathematical Economics, Macmillan, New York.

[19] W. Leontief (1966), Input-Output Economics, Oxford University Press, New York.

[20] T.C. Miller, T.L. Friesz, and R.L. Tobin (1996), Equilibrium Facility Location in Networks, Springer-Verlag, Berlin.

[21] M. Morishima (1964), Equilibrium, Stability and Growth, Clarendon Press, Oxford.

[22] H. Moulin (1981), Théorie des Jeux pour l'Économie et la Politique, Hermann, Paris.

[23] A. Nagurney (1999), Network Economics: A Variational Inequality Approach, Kluwer, Dordrecht.

[24] J. von Neumann and O. Morgenstern (1953), Game Theory and Economic Behavior, Princeton University Press, Princeton.

[25] H. Nikaido (1968), Convex Structures and Economic Theory, Academic Press, New York.

[26] K. Okuguchi and F. Szidarovszky (1990), The Theory of Oligopoly with Multi-product Firms, Springer-Verlag, Berlin.

[27] J.M. Ortega and W.C. Rheinboldt (1970), Iterative Solution of Nonlinear Equations in Several Variables, Academic Press, New York.

[28] M. Patriksson (1994), The Traffic Assignment Problem: Models and Methods, VSP, Utrecht.

[29] M. Patriksson (1999), Nonlinear Programming and Variational Inequality Problems: A Unified Approach, Kluwer, Dordrecht.

[30] V.M. Polterovich (1990), Equilibrium and Economic Mechanism, Nauka, Moscow (in Russian).

[31] B.T. Polyak (1983), Introduction to Optimization, Nauka, Moscow; English translation in Optimization Software, New York, 1987.

[32] P.A. Samuelson (1983), Foundations of Economic Analysis, Harvard University Press, Cambridge.

[33] H.E. Scarf and T. Hansen (1973), Computation of Economic Equilibria, Yale University Press, New Haven.

[34] A. Schrijver (1986), Theory of Linear and Integer Programming, Wiley, New York.

[35] A.N. Tikhonov and V.Ya. Arsenin (1977), Solutions of Ill-Posed Problems, John Wiley and Sons, New York.

[36] L. Walras (1874), Eléments d'Économie Politique Pure, L. Corbaz, Lausanne; English translation: Elements of Pure Economics, Allen and Unwin, London, 1954.

Index

Mathematics in Science and Engineering
Edited by C.K. Chui, Stanford University

Recent titles:
C. De Coster and P. Habets, *Two-Point Boundary Value Problems: Lower and Upper Solutions*
Wei-Bin Zang, *Discrete Dynamical Systems, Bifurcations and Chaos in Economics*
I. Podlubny, *Fractional Differential Equations*
E. Castillo, A. Iglesias, R. Ruíz-Cobo, *Functional Equations in Applied Sciences*
V. Hutson, J.S. Pym, M.J. Cloud, *Applications of Functional Analysis and Operator Theory (Second Edition)*
V. Lakshmikantham and S.K. Sen, *Computational Error and Complexity in Science and Engineering*
T.A. Burton, V*olterra Integral and Differential Equations (Second Edition)*
E.N. Chukwu, *A Mathematical Treatment of Economic Cooperation and Competition Among Nations: with Nigeria, USA, UK, China and Middle East Examples*
V.V. Ivanov and N. Ivanova, *Mathematical Models of the Cell and Cell Associated Objects*
Z. Zong, *Information-Theoretic Methods for Estimating Complicated Probability Distributions*
A.G. Ramm, *Dynamical Systems Method for Solving Operator Equations*
J. Mishra and S.N. Mishra, *L-System Fractals*

www.ingramcontent.com/pod-product-compliance
Lightning Source LLC
LaVergne TN
LVHW011756200926
848484LV00003B/721
* 9 7 8 0 4 4 4 5 3 0 3 0 1 *